CONTR

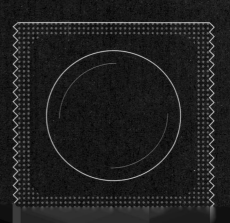

The MIT Press Essential Knowledge Series

A complete list of the titles in this series appears at the back of this book.

CONTRACEPTION

A CONCISE HISTORY

DONNA J. DRUCKER

The MIT Press | Cambridge, Massachusetts | London, England

This book was set in Chaparral Pro by Toppan Best-set Premedia Limited. Printed and bound in the United States of America.

Library of Congress Cataloging-in-Publication Data

Names: Drucker, Donna J., author.
Title: Contraception : a concise history / Donna J. Drucker.
Other titles: MIT Press essential knowledge series.
Description: Cambridge : The MIT Press, [2020] | Series: The MIT Press
    essential knowledge series | Includes bibliographical references and index.
Identifiers: LCCN 2019024256| ISBN 9780262538428 (paperback) | ISBN
    9780262357579 (ebook)
Subjects: | MESH: Contraception—history
Classification: LCC RG136.2 | NLM WP 11.1 | DDC 618.1/8—dc23 LC
    record available at https://lccn.loc.gov/2019024256

10   9   8   7   6   5   4   3   2   1

CONTENTS

Series Foreword   vii
Preface   ix
Acknowledgments   xi

1   Why Contraception?   1
2   Contraception before the Pill   15
3   The Pill and Its Successors   65
4   Nonhormonal Contraception after the Pill   91
5   Contraception in the Reproductive Justice
    Framework   133
6   The Future of Contraception   157

Glossary   183
Notes   187
Bibliography   209
Further Reading   231
Index   233

SERIES FOREWORD

The MIT Press Essential Knowledge series offers accessible, concise, beautifully produced pocket-size books on topics of current interest. Written by leading thinkers, the books in this series deliver expert overviews of subjects that range from the cultural and the historical to the scientific and the technical.

In today's era of instant information gratification, we have ready access to opinions, rationalizations, and superficial descriptions. Much harder to come by is the foundational knowledge that informs a principled understanding of the world. Essential Knowledge books fill that need. Synthesizing specialized subject matter for nonspecialists and engaging critical topics through fundamentals, each of these compact volumes offers readers a point of access to complex ideas.

Bruce Tidor
Professor of Biological Engineering and Computer Science
Massachusetts Institute of Technology

This book concerns mechanical, chemical, pharmaceutical, and behavioral methods intended to prevent conception—the period before a fertilized egg is implanted in the uterus, or approximately one to two weeks following a sexual encounter involving ejaculation. The line between contraception and regulation for amenorrhea (absence of a monthly period) has not always been clear, especially before medical science confirmed the timing of ovulation in the female reproductive cycle in the 1920s. The termination of a pregnancy before the movement of the fetus, known in the past as "quickening," was often considered an aspect of menstrual regulation instead of an abortion. Thus, some of the herbal methods described in chapter 2 may have served as menstrual regulators or early abortive medications instead of contraceptives.

Though abortion following an established pregnancy is a key element of reproductive health and justice, and part of a continuum of processes for ending pregnancies, as a post-implantation technological process, this book does not cover it. Given its vast medical, historical, and legal complexities, it deserves a book of its own.

This book, like others in the field of reproductive history, must balance the relevance of "women" and "men" as categories of personhood with specific historical meaning and the necessary inclusion of transgender and nonbinary individuals, whose ability to become pregnant or to impregnate someone else may not correlate to their gender identity. I use gendered terms when needed to discuss particular experiences of self-identified women and men in the past and use neutral terms as much as possible.

The term *contraception* as opposed to *birth control*, which could refer to any technology or technological process used between conception and live birth, is used throughout. The term *birth control* is used only if individuals or organizations themselves used it in historical context.

Finally, this book is not a guide to choosing a contraceptive device or practice in the present. Please consult a health-care practitioner for advice.

ACKNOWLEDGMENTS

Though this book has one name on the front cover, many people helped bring it into being.

Katie Helke, acquisitions editor at the MIT Press, first suggested that I write for this series at the 2017 Society for the History of Technology conference in Philadelphia, Pennsylvania. I am grateful for her encouragement.

I thank Heather Munro Prescott and the two anonymous readers of an earlier version for their helpful suggestions. Thanks also to the Lemelson Center for the Study of Invention and Innovation at the Smithsonian National Museum of American History for a 2017 Travel-to-Collections Award, which led to finding the Parke, Davis catalogs and many of this book's illustrations. Jim Roan, the museum's reference librarian, and Diane Wendt, deputy chair and associate curator of the museum's Division of Medicine and Science, were generous with their time and advice. I am likewise grateful to the New England Research Fellowship Consortium for a travel grant that supported the discovery of additional sources at Smith College, the Center for the History of Medicine at the Francis A. Countway Library of Medicine, and the Arthur and Elizabeth Schlesinger Library on the History of Women in America.

This book benefitted from conversations with Laura Kelly, Jesse Olszynko-Gryn, and the audience at the Reproductive and Sexual Health Activism, c. 1960–Present Workshop in Glasgow, Scotland, in July 2018. I also thank Jessica Borge, Micheline Egan, Prince Guma, Alana Harris, Rebecca Hodes, Claire L. Jones, Miriam Klemm, Kate Law, Carolyn Herbst Lewis, Liz McMahon, Alejandra Osorio, Lucia Pozzi, Caroline Rusterholz, and Laura Ann Twagira for sharing their work, source ideas, and thoughts with me. The thoughtful hospitality of Paul Maddern at the River Mill Writer's Retreat in Northern Ireland sustained me in a period of intense composition.

My friends and family members Tobias Boll, Michele Campbell, Kate Costello, Michelle Cunningham-Wandel, Stefan Glatzl, Kyla Jemison, Corinna Norrick-Rühl, Clark A. Pomerleau, Charles Peters, Lora Stephan, and Katie Watson, along with my supervisor, Christoph Merkelbach, supported me throughout the research and writing of this book. I could not ask for a more loving set of parents and relatives than Donald S. and Diane K. Drucker, Alan and Adrienne Drucker, and Charles and Betty Watson.

This book is dedicated to the memory of Mark A. Price, my best bud.

WHY CONTRACEPTION?

Controlling, timing, or avoiding pregnancy is a concern for anyone involved in sexual activity with the potential for sperm-egg contact. The history of contraception is important because controlling birth, either to guard against or to promote pregnancy, has been a concern throughout recorded human history.[1] Learning such history sheds light on the scientists, manufacturers, government officials, distributors, salespersons, and activists who paved the way for the variety of contraceptive technologies used today. Additionally, this book provides readers historical context for their own reproductive lives, contraceptive use, and decision-making processes.

More broadly, this book also frames the history of contraception in a wider context of population control, eugenics (including involuntary sterilization), racist and classist restrictions on birth control access, and the extent to which people do or do not accept technological methods into their sexual and reproductive lives. Various technological methods can be embraced or rejected for a variety of reasons, including mental health (loss of libido or desire), physical health (increased bleeding or spotting), and allergies (such as to latex). Additionally, those with strict religious or moral beliefs, such as those who adhere to Roman Catholicism's prohibition of technological contraceptives and those who avoid hormonal or technological modifications to the body, both favor timing methods, which can also include withdrawal. Some of these individuals, however, may accept the use of external technologies, such as a thermometer or fertility computer, in order to avoid more invasive or morally objectionable internal technologies. Others may use technologies such as sex toys or dolls, with or without the presence of a partner, in order to avoid sperm-egg contact completely. Studying the reasoning behind the use or nonuse of contraceptive methods thus illuminates broader themes in the history of human-technological interaction.

This book raises broader questions not only about the relationship of individuals to technologies but also about the ways that contraceptives play a role in local, national, and international politics. Laws and policies from the US's Comstock Act to Ireland's Criminal Law Amendment Act affected and continue to affect people's personal lives, livelihoods, and decision making. Laws and regulations govern the intellectual property of contraceptives (such as the chemical composition of spermicides); manufacturing standards; testing on animal and human subjects; legal requirements concerning advertising, sales, and distribution; and the parameters under which sales are allowed, such as age restrictions or prescription requirements. National or state policies can force people into involuntary sterilization, unwanted intrauterine devices, or hormonal implants in the service of "population control," as happened during the 1975 to 1977 Emergency in India. The presence or absence of legal contraceptives in a state or nation serves as a symbol of its commitment to women's and human rights—in other words, its commitment to reproductive justice.

This book is a history of contraceptive technologies from the opening of the first birth control clinic in Amsterdam, the Netherlands, in 1882 to the present. It traces the research, development, manufacturing, distribution, and use of contraceptive methods that were and are marketed and sold to the public. Those methods were and are available with or without a prescription, for people of all genders. The history of contraception involves the synthesizing of diverse histories, including the history of technology, women's and gender history, the history of sex and reproduction, population control studies, legal history, and political history. It requires a broad understanding of individual behavior and identity formation;

nonprofit advocacy groups and independently wealthy individual advocates; religious organizations; governmental policies at various levels and the execution thereof; and technological development, manufacturing, and distribution, among other factors. Geographically, the primary emphases in this book are on the United States and Western Europe, with secondary emphases on the Caribbean, Peru, Eastern Europe, sub-Saharan Africa, India, and Japan. This book draws on existing scholarship in four areas: first, chronologically and temporally restricted histories of contraception; second, histories of sexuality and sexology; third, histories of fertility and infertility; and fourth, histories and theories of feminist health and reproductive justice. It is organized both chronologically and according to the type of technology under development.

This book identifies the opening of Aletta Jacobs's birth control clinic as the beginning of a modern contraceptive era. Of course, condoms, behavioral methods (abstinence and withdrawal), and herbal preparations existed long before 1882. The founding of her clinic, however, marks the beginning of organized, internationalized, and systemic approaches to contraception based on the idea that professional medicine had to address women's need for contraception—and that rubber, chemical, and pharmaceutical industries should, too. The clinical provision of contraception marked a moment when a physician started to take seriously the prevention of pregnancy as a medical concern, and a moment when a doctor prescribed the use of a technology for a patient who was not ill. When word spread about Jacobs's clinic, it sparked thinking about how, when, and why pregnancies could be spaced or prevented—and by whom. In short, "control over the timing, means ('artificial' or 'natural'), and frequency of conception, and especially its prevention, was at the heart of the modernist reproductive project."[2]

Starting with the establishment of Jacobs's clinic in Amsterdam makes sense for more specific reasons as well: (1) the clinic made the Mensinga diaphragm available to women without the approval or knowledge of their husbands; (2) her clinic established a medical service model that American and British birth control clinics would follow; and (3) Margaret Sanger highlighted her attempted contact with Jacobs (and her successful contacts with Dutch male physicians) publicly as a signifier of her own determination and expertise. Thus, the opening of the Jacobs clinic, while providing contraceptives to a relatively small number of women until 1894, carries ongoing symbolic weight in the narrative of contraceptive history that advocates (particularly Sanger and the English advocate Marie C. Stopes) told. The year 1882 also marks a clear moment in the history of both technology and medicine: the Mensinga diaphragm was a woman-controlled method that was fitted and distributed by a female physician. It was a technology marked openly as one that women could prescribe, control, and use without male interference.

This book examines the history of contraceptives from the perspectives of reproductive justice and feminist technological studies. It argues that tracing access to, research and development of, and use of contraceptive technologies is an outward measure of how a society values human selfhood and autonomy. A reproductive justice approach encourages examination of contraceptives on a three-pronged sociopolitical level: does it permit anyone capable of pregnancy to have children, not to have children, and to raise children in a safe and healthy environment?[3] At the same time, a feminist technology studies approach points toward the study of contraceptives as material artifacts on an individual level: in evaluating a technology from a feminist perspective, does it facilitate or constrain equitable gender relations?[4] Contraceptive knowledges, materials, and practices that meet people's changing needs over time both support and strengthen efforts for both reproductive justice and feminist technology.

Themes

This history of contraception addresses four themes that resonate across the last 140 years of contraceptive modernity under examination here. The first theme is power relationships: who in a sexual relationship, however short- or long-term, wants and can use contraception; who has legal access to what kind of technology; what the quality of the technology is; how affordable it is; how to use it properly; and when, where, and how often it is available. Gendered power inequalities in a relationship often determine contraceptive use and efficacy, as do the legal structures and the social milieu in which those relationships operate. Social and legal inequalities across countries and cultures permitted men historically to have sex with their wives with or without the latter's consent and for the husband to throw out contraception or to be violent against her if the wife tried to use it without his knowledge or consent. Furthermore, men have opposed contraception in order to demonstrate virility by impregnating as many women in or outside of marriage as they could (ideally producing more sons), whether or not further pregnancies would harm the woman and whether or not they could afford to feed, clothe, and house the resulting children.[5] More equality in marital and sexual relationships, especially after the second wave of feminism in the last third of the twentieth century, along with changes in laws and social mores, greatly affected women's ability to exercise their right to choose and to use contraception within and outside marital relationships.

The second theme is the persistence of certain contraceptive methods over time. Some methods appear at one point in the historical record and then reappear in different forms or are later repackaged to attract a new generation of users. While understandings of how methods work may change as science advances, with or without that precise understanding, some of them have remarkable staying power. The persistence of methods may speak to their efficacy, ease of use, adaptability to different situations and cultures, and complementarities in the motives of users, manufactures, and promoters. For example, methods that depend on the timing of sex according to a woman's menstrual cycle (and to some extent sexual positioning and the length of a sexual encounter) have been a longtime part of the contraceptive repertoire of many cultures. However, in February 2017, a mobile app based on fertility timing called Natural Cycles was approved in Europe, and in August 2018, the United States Food and Drug Administration (FDA) approved the technology as well. An ongoing interest in nonhormonal methods for contraception—based on religious beliefs, a desire to avoid the pill, and an overall lifestyle favoring "natural" approaches to living—were all impetuses for the FDA to approve the app with its accompanying thermometer (chapter 4).[6] The science behind Natural Cycles is not new, but the mobile apparatus that supports it is.

Third, technology that works for one acceptable purpose may also be used for contraception, though the contraceptive purpose may be legally or unofficially prohibited.

For example, homemade and commercial vaginal douches could serve a contraceptive purpose, but in places where contraceptives were illegal in the early twentieth century, they were most often marketed and sold as cleansers and deodorizers. Women could read in between the lines of the advertisements, experiment, or get advice from friends or family to find out that douches could work (albeit with limited effectiveness) as contraceptives. Contraception is often a "camouflage technology": its real purpose hidden in plain sight.[7]

Fourth, it is clear that contraception is never a neutral technology. Contraception can be used for any number of reasons. It can be temporary or permanent, involuntary or voluntary, freely chosen or enforced by a government or an intimate partner, enhance sexual pleasure or take away from it. For example, governments may force certain populations to undergo sterilization, particularly for eugenic purposes, if they deem those populations to be somehow "unfit" to continue reproducing. Those targeted can include poor people, mentally or physically disabled people, and people who have committed sex crimes or other offenses, among others. On the other hand, wealthy, light-skinned, and high-status women whose reproduction was deemed socially valuable have often had difficulties getting sterilized voluntarily until they had produced a certain number of children. Manufacturing of, access to, distribution of, and use of contraception varied and

Contraception is never a neutral technology.

continue to vary widely in availability, purpose, and intent. Having multiple options for one's choice of contraceptive method across a reproductive lifetime—options that are affordable, steadily accessible, unproblematic to use, effective, and safe—is rare indeed.

Organization

This book is organized chronologically, beginning in the 1880s when the diaphragm was first obtainable in Jacobs's Amsterdam clinic. Chapter 2 outlines contraceptives that were available between the 1880s and 1960, when the US FDA legalized the pill for contraceptive use. These include timing and behavioral methods, such as withdrawal and periodic abstinence; chemical and herbal methods, such as commercial douches and spermicidal jellies; and barrier methods, such as the diaphragm and condom. It uses excerpts from literature, oral histories, and memoirs to shed intimate light on the role played by these methods and devices in people's everyday lives. Some of these methods are unsafe and ineffective by contemporary standards and have fallen into disuse, while the material technology undergirding others has improved and remains sound in the present.

Chapter 3 looks at how the pill and its chemical descendants were developed and then spread across the world—sometimes in the name of providing women with agency over their reproductive lives and sometimes with the intention of population control. It traces the original testing, research, and development of the first hormonal methods in Mexico, the continental United States, and Puerto Rico. As more pharmaceutical companies within and beyond the US began quickly to manufacture the pill, it altered the state of sexual and marital relations according to existing laws, gender ideals, and moral standards. After the chemistry behind the pill was widely understood, pharmaceutical companies began to experiment with other forms of delivery and adapted hormone formulations originally created for other purposes to contraceptive ones. One of the most controversial hormonal methods, Depo-Provera, played two distinct roles in contraceptive history. First, in the US context, its parent company Upjohn fought a decades-long battle against feminist health organizations and the FDA to get it approved for contraceptive use in 1992.[8] Second, in the context of the developing world, countries interested in population control—and sometimes with eugenic intentions—made it available (often for free) to poor, racial and ethnic minority, and disabled individuals in the 1980s without providing them full information on the shot's side effects. Research and development into the "male pill" closes the chapter.

Chapter 4 addresses technological changes in the non-hormonal contraceptive methods first described in chapter 2—how materials, chemicals, and forms of delivery changed along with the ways that decades-old technologies reemerged to provide users with new options. For example, it reviews how the cervical cap, a method little-used outside Germany and England, reemerged in the 1970s US as an alternative to the pill, thanks to the feminist women's health movement. Also, it investigates how the Roman Catholic Church's 1968 authorization of a timing method alone as the only non-sinful method of contraception was a critical moment in the technology's history: it reinvigorated interest in timing methods, caused some to leave their faith, and pushed others to embrace the pill nonetheless.

Chapter 5 introduces the concept of reproductive justice, a simple yet robust theory that structures worldwide reproductive activism in the twenty-first century. Reproductive justice consists of the three basic principles mentioned above—the right to have a child, the right not to have a child, and the right to parent children safely and healthily. Reproductive justice, based as it is in human rights principles, provides a foundation for activism at local, national, and international levels. It likewise provides the groundwork for analyzing reproductive injustices in the past and a vision for the future in which everyone can decide to become a parent or not in environments that support the human flourishing of all. Access to safe contraception is thus one element of a reproductively just world.

Chapter 6 is divided into three parts: first, its focus is on the ways that existing contraceptives must adapt in order that all people, including overweight or obese individuals and trans individuals, can use them safely and effectively. The second section covers present-day contraceptive technologies and new forms of distributing them, such as brothels with robot-only sex workers, where people can have sex with inanimate, humanlike machines that cannot become pregnant or transmit sexually transmitted infections. It ends with an examination of the current world situation regarding contraception and the efforts of international health organizations and pharmaceutical companies to support contraceptive access, especially to the approximately 225 million women who are deprived of it.[9] There is a long way to go to ensure that contraceptive information and methods are accessible to everyone. As long as humans have reproductive sex, people will continue to innovate new ways to manage conception and contraception alike.

CONTRACEPTION BEFORE THE PILL

Modern contraception has its roots in neo-Malthusianism. Thomas Robert Malthus, in the book *An Essay on the Principle of Population* (1798), theorized that while the availability of resources to feed humans grows at a steady state, the population expands at a higher rate than those resources. Periodic food shortages and the deaths by starvation that resulted were nature's way of correlating population growth with resources. War and disease had similar effects, which Malthus called "positive checks," and these fell in line with "preventative checks," or individual decisions to practice abstinence, to delay marriage, or to use other means (such as withdrawal) to curb the number of children. The book, as is abundantly clear, justified a lack of care toward poor and indigent people, and it contained the seeds of the doctrine—neo-Malthusianism—that motivated many contraceptive promoters in the late nineteenth and early twentieth centuries.[1]

Advocates of neo-Malthusianism left behind Malthus's cruel idea of positive checks on population but developed a new perspective on preventative checks that justified the distribution of contraceptive information and goods. For them, preventative checks were actions that individuals could take to limit offspring and to forward the aim of "population control." In the late nineteenth century, this term signified the goal of maintaining the human population at a steady level, limiting interclass conflict, and helping people not to produce more children than they wanted or could afford. It is not hard to see how those goals for using contraceptive technologies—which were often classist, racist, and paternalistic—morphed into arguments for using those same technologies for eugenic purposes. And rarely did population controllers mention a connection between freedom from pregnancy concerns and an increase in women's sexual pleasure and satisfaction.

Thus, although the public discussion, manufacture, marketing, sale, and distribution of goods for contraceptive purposes were illegal in many countries until the late 1920s or 1930s or even later, those goods and knowledge of behavioral methods were nonetheless obtainable depending on multiple factors—access to the goods themselves, funding to purchase them, knowledge of how to use them properly (often requiring literacy or access to healers or medical professionals), motivation to use them, agreement of a partner, and often willingness to break the law. The following sections review each of the methods in use between the 1880s and 1960, some of which were free and available to anyone, and others that were available only for purchase.

Diaphragms and Cervical Caps

The first mention of custom-made rubber cervical caps was in the Berlin-based physician Friedrich Adolph Wilde's 1838 treatise "Das weibliche Gebär-unvermögen" ("The Female Inability to Give Birth"), but dependable and commercially available contraceptive barrier methods for women were developed after the vulcanization of rubber in the United States and England in 1844.[2] Vulcanization made rubber stronger, more heat resistant, and more elastic, and following its introduction in the world market for industrial purposes, rubber goods manufacturers also realized its utility for medical devices. Rubber pessaries marketed as "womb supporters" were available in the US in the early 1860s for women with gynecological problems such as prolapsed or tipped (tilted) uteri, although they were not effective as contraception, even in an unspoken fashion, due to the presence of a hole in the middle to relieve pressure on the vagina.

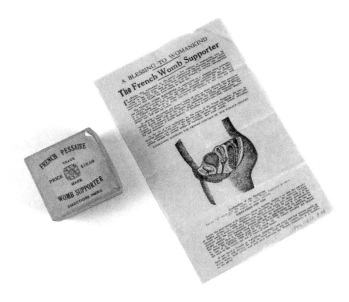

Figure 1 French Pessaire Womb Supporter pessary with instructional flyer, ca. 1880s. *Source:* Courtesy of Medicine and Science Collections, National Museum of American History, Smithsonian Institution, Washington, DC.

The contraceptive technology that the Dutch physician Aletta Jacobs provided beginning in January 1882 was a Mensinga diaphragm, also known as an occlusive pessary—a soft rubber barrier held fast in the vagina over the cervix by a flat spring. It was named after its inventor, the German physician W. P. J. Mensinga, who was first active in Leipzig, then later in Flensburg and Breslau. The device, which eventually came in five different sizes, required initial fitting by a physician, but the woman herself could handle insertion and removal with her fingers or an inserter shaped like a narrow shoehorn. In the central German state of Thuringia, obtaining contraception in the 1880s was popularly known as "going to Flensburg."[3] Knowledge of Mensinga's work spread across northern Europe through word of mouth and through his pseudonymous publications. Jacobs came across Mensinga's diaphragms while researching contraception on her own and likely obtained her first supply of diaphragms directly from him.[4]

The Mensinga diaphragm (later known as a Dutch cap) and knowledge of Jacobs's clinic entered the American market through the writing and efforts of the activist Margaret Sanger. Although diaphragms were illegal in the US under the Comstock Act of 1873, which prohibited the manufacture, distribution, and sale of obscene goods, Sanger intended to find a way to distribute them. She wanted to learn diaphragm fitting from Jacobs herself, but Jacobs, who had closed her clinic in 1894, refused to see her when she visited in February 1915, citing Sanger's lack of a medical degree. Sanger learned the technique from another member of the Dutch Neo-Malthusian League, Johannes Rutgers, and her public career as a birth controller began in October 1916 with the short-lived Brownsville Clinic in Brooklyn, New York.[5] Sanger asked wealthy society women such as Katharine McCormick to

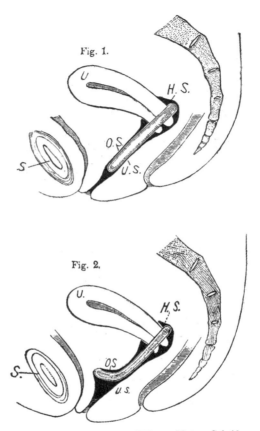

Figure 2 German occlusive pessary with metal inserter. In Matrisalus [pseud.], *Den Frauen Schutz!* [*For the Protection of Women*] (Leipzig, 1898). *Source:* Courtesy Center for the History of Medicine, Francis A. Countway Library, Harvard University, Boston, MA.

Fig. 3.

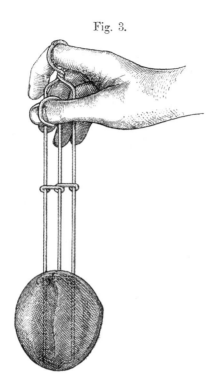

Figure 2 (continued)

smuggle diaphragms into the US illegally (chapter 3), and she herself smuggled them in through Canada, but this transportation method was too irregular and risky to depend on. Sanger, with her second husband J. Noah Slee, financed the Holland-Rantos Company in 1925 to produce diaphragms, cervical caps, and other rubber goods for the clinics that were spreading rapidly across the country.[6]

Rubber goods manufacturers in the United States and other countries were aware of the market for diaphragms, cervical caps, and condoms and aimed to capture some of the profits for themselves. In addition to the diaphragm, these firms also manufactured cervical caps, which were smaller and also made of rubber and fit directly over the cervix. Julius Schmid, a successful condom manufacturer, started manufacturing diaphragms in 1923, beating Holland-Rantos to the US market by two years. The English advocate Marie Stopes established a firm that produced what she named the "Racial" line of contraceptive technologies, including "Pro-Race" cervical caps, diaphragms, and sponges. She named these devices after her aim to improve the human race through the distribution of contraceptives to poor and indigent women. The Australian gynecologist Norman Haire, who worked for many years in England, collaborated with the London-based company Lamberts (Dalston) Ltd. in 1922 to produce a cap that he publicly endorsed in print advertisements.[7]

Contraceptive advocates aimed to make diaphragms and caps as widely available as possible. In Germany, diaphragms and spermicides could be ordered through the magazine *Sexual Hygiene*. In Vienna, both municipally funded and private clinics offered advice, condoms, and cervical cap fittings. In Japan, diaphragms became available in clinics after Margaret Sanger visited Tokyo in February and March 1922. And private doctors could prescribe diaphragms for women in South Africa, have them shipped from England, or obtain them from clinics that opened starting in 1932. Word about the Cape Town clinic soon reached Namibia, and Namibians wrote to its staff for information and supplies. That clinic likewise established a contraceptive mail-order service for rural women that continued until World War II, and it purchased tram tickets for poor city women who could not afford the journey to the clinic on their own.[8] On a visit to India in 1936 and 1937, the English feminist and contraceptive promoter Edith How-Martyn observed doctors fitting diaphragms in clinics and chemists' shops, and she worked with English manufacturers like Prentif to ship contraceptive supplies there. Given their expense, however, diaphragms and caps were available only to middle- and upper-income women, and attempts by both Sanger and Stopes to establish a market for their respective brands of contraceptives in India failed.[9]

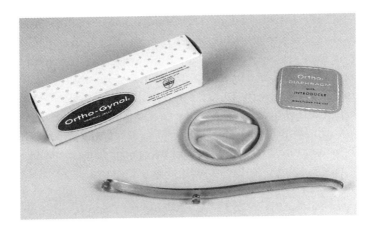

Figure 3 Ortho-Gynol diaphragm set including applicator and instruction booklet, ca. 1940s. *Source:* Courtesy Medicine and Science Collections, National Museum of American History, Smithsonian Institution, Washington, DC.

Although the diaphragm was the device that birth control clinics most often recommended to their clients in the 1920s and 1930s, largely because it put contraceptive control in women's hands, it was far from problem-free. In *Control of Contraception: A Clinical Medical Manual* Robert Latou Dickinson included images of all the different ways a diaphragm could fail: the physician could fit a woman with a device that was too big or too small, the woman could place it incorrectly, or it could become dislodged during vigorous or prolonged intercourse.[10] It was less effective if used without an accompanying spermicide; it was messy and smelly; it needed to be left in for several hours after use, then washed and dried; and proper insertion, removal, cleaning, and storage required the running water and privacy that many poor women simply did not have.[11] It was unavailable to American women without a medical precondition until the *United States v. One Package of Japanese Pessaries* appeals court decision in December 1936.[12] Moreover, visiting a birth control clinic for a diaphragm or a cervical cap remained fraught for women who were new to thinking about sex and fertility and who were breaking laws or religious or cultural taboos. Being fitted for a diaphragm was often an exercise in embarrassment. Mary McCarthy's 1954 novel *The Group* depicts a new college graduate's experience of a diaphragm fitting in 1933:

> Dottie did not mind the pelvic examination or the fitting. Her bad moment came when she was learning how to insert the pessary by herself. ... As she was trying to fold the pessary, the slippery thing, all covered with jelly, jumped out of her grasp and shot across the room and hit the sterilizer. Dottie could have died. But apparently this was nothing new to the doctor and the nurse. "Try again, Dorothy," said the doctor calmly, selecting another diaphragm

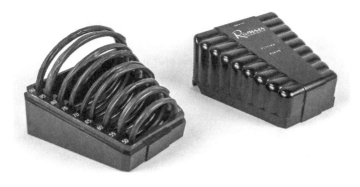

Figure 4 Ramses diaphragm-fitting kit for physicians, Julius Schmid, New York, NY, ca. 1940s. *Source:* Courtesy Medicine and Science Collections, National Museum of American History, Smithsonian Institution, Washington, DC.

of the correct size from the drawer. And, as though to provide a distraction, she went on to give a little lecture on the history of the pessary, while watching Dottie's struggles out of the corner of her eye.[13]

The Irish writer and later feminist activist June Levine had a similar experience when she visited a Dublin doctor around 1949 to have an illegal diaphragm fitted. Her memoir highlighted the precariousness of access to the device: after the doctor gave it to her, he said, "Now, I'll give you this, but if you know anyone going to England or Belfast tell them to bring one and give it back to me. Size 95 [millimeters]. It's the only way I can get them, you know." So even physicians willing to break the law were unable to procure the actual goods that would help their patients regularly. Levine expressed her ongoing frustration with the device and described its effects on her sexual desire: "I doubt that I ever got the hang of that 'thing,' hitting some part of the bathroom with it every time it sprang from between my thumb and forefinger. ... Scrambling around the bathroom floor in search of the escapee, fingers sticky and smelly from the cream, at best cooled my ardour, at worst put me in a vicious rage."[14]

Thus, the diaphragm and cervical cap largely remained the province of wealthier women in Western countries with access to ongoing professional medical care. It was effective enough for those who could get it fitted professionally and clean and use it properly, but it was far from ideal, and did not provide protection against sexually transmitted infections. While some women may have found relief in using a barrier method under their own control, women like Levine expressed how it dampened sexual desire.

Spermicides, Sponges, Suppositories, and Douches

Once diaphragms and cervical caps became a regularized part of Westernized clinic offerings, most health-care professionals recommended that they be used with an additional spermicide or jelly—in short, pairing a barrier method with a chemical method brought together two different approaches to contraception that would improve efficacy overall. The barrier-chemical combination has its roots in the work of the freethinking Massachusetts-based physician Charles Knowlton, who recommended in his 1832 pamphlet that women place a small piece of natural sponge moistened with water in their vaginas, tied with a ribbon for easy removal. Following intercourse, the woman should then use a syringe to douche with a "solution of sulphate of zinc, of alum, pear-ash, or any salt that acts chemically on the semen." If no chemicals were at hand, Knowlton recommended douching with plain water. This sponge-and-douche method "not only dislodges the semen pretty effectually, but at the same time destroys the fecundating property of the whole of it."[15] A second edition of the pamphlet landed Knowlton in prison for three months, a fate similar to that of the birth control campaigners Charles Bradlaugh and Annie Besant, who republished Knowlton's *Fruits of Philosophy* in England in 1877. As a result of Bradlaugh's and Besant's actions, 185,000 copies of the text were published between 1877 and their 1881 trial, and a new generation of English readers learned, and perhaps practiced, Knowlton's sponge-and-douche method.[16]

Both douches and suppositories—plugs of a sticky substance designed to melt in the vagina and to have both barrier and chemical spermicidal qualities—contained ingredients that were often available at home or over-the-counter at drugstores. In 1885, the English pharmacist Walter Rendell produced the first commercial suppository, made from cocoa butter and quinine sulfate. Suppositories had their own problems, however, because they melted inside the vagina unevenly, provided incomplete coverage, and usually leaked out during and after intercourse, leaving an unpleasant mess.[17]

The active ingredients in douches and suppositories are remarkably similar from the 1830s through the 1930s. Common ingredients in commercial douches included weak acids and astringents, such as copper water, baking soda, vinegar, carbolic acid, bichloride of mercury, soap suds mixed with coconut oil, borax, alum, citric acid, and salt. The popular brand Lysol used creosote—a harsh germ killer that was a distillate of wood and coal—as its active ingredient. Zonite used sodium hypochlorite, first discovered as a wound cleanser. One could also soak a wad of cotton, rubber sponge, or sea sponge with these substances to use as a barrier preventative.[18] Additional common active ingredients in suppositories included boric acid, quinine, or salicylic acid. Most of these concoctions did little more than irritate women's vaginal tissue, but some, like quinine, caused dizziness and headaches.

Some birth controllers endorsed homemade remedies, especially for poor and rural women. Margaret Sanger's

Figure 5 DeWitt's Hygienic Powder cannister, ca. 1906. *Source:* Courtesy Medicine and Science Collections, National Museum of American History, Smithsonian Institution, Washington, DC.

first pamphlet, *Family Limitation* (1914), included recipes for simple astringent douches and cocoa butter–based suppositories.[19] Women in Tamil-speaking southern India used neem oil, honey, and scooped-out lime halves as barrier methods and prepared douches from alum, neem leaf tea, neem toothpaste, tamarind tea, and vinegar.[20] The Jamaican Birth Control League provided "a pamphlet for women in the countryside explaining a range of home methods (from douches to homemade sponges made from cotton wool and oil), but clearly stated that these methods were 'not so sure as doctor's methods,' advising women to get professionally fitted for a diaphragm." Marie Stopes continued to promote the cotton wool and oil method for poor and uneducated women in India as late as 1952, even though she did not believe in its efficacy.[21]

In the late nineteenth and early twentieth centuries, commercial douches came onto the US, German, and British colonial markets and were advertised as providing contraceptive security that was superior to homemade concoctions. However, because there were no government regulations for the dozens of products that existed, their efficacy was not guaranteed. In a 1933 report from the Newark Maternal Health Center in New Jersey, 91.5 percent of 1,978 women had tried at least one contraceptive method, 80 percent had tried multiple methods, 507 had douched with Lysol, and 239 had douched with plain water, with limited success.[22] Commercial douches were no

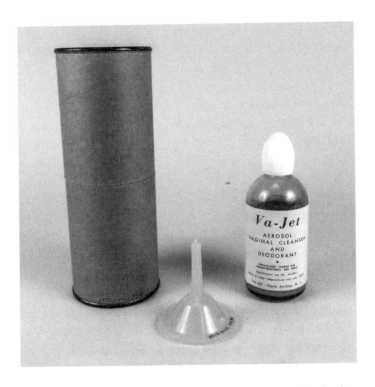

Figure 6 Va-Jet Aerosol Vaginal Cleanser and Deodorant, ca. 1920s, Perth Amboy, NJ. *Source:* Courtesy Medicine and Science Collections, National Museum of American History, Smithsonian Institution, Washington, DC.

more than vaginal deodorant sprays with no contraceptive function and often caused pain and irritation due to their strong astringency.

Birth controllers and pharmaceutical manufacturers also experimented with manufacturing, distributing, and selling spermicidal foam powders and tablets. In the 1930s, the Oxford University zoologist John Rendell Baker favored a mercury derivative as an active ingredient in his Volpar (Voluntary Parenthood) tablets, which were later found to be toxic.[23] A German brand of foaming tablets, Speton, were mentioned in a 1930 article in the journal *Estonian Physician*, and sponges and foam powders were also available for free in the first Bermudian birth control clinic in 1937. Two Florida-based inventors in the 1930s created a foaming powder with sodium lauryl sulfate (a common ingredient in soaps and shampoos) called Duponol that Margaret Sanger had shipped to India and China in 1937 and 1938 as a potentially cheaper diaphragm alternative. After World War II and the partition of India in 1947, a new generation of birth control organizations, including the Rockefeller Foundation–funded Population Council, became interested in developing contraceptives specifically for women living in hot climates. The Population Council underwrote the Khanna Study (1953–1960) on the Durafoam contraceptive tablet in the Indian state of Punjab, but the researchers' lack of understanding of local mores, health needs, and existing patterns of female

Figure 7   Smith's Contab Contraceptive Foam Tablets, Smith, Stanistreet & Co., Ltd., Calcutta, India, ca. 1940s. *Source:* Courtesy Medicine and Science Collections, National Museum of American History, Smithsonian Institution, Washington, DC.

Figure 8 Spermet contraceptive vaginal tablets, Haifa, Israel, ca. 1940s. *Source:* Courtesy Medicine and Science Collections, National Museum of American History, Smithsonian Institution, Washington, DC.

infanticide doomed the study from the start.[24] Regardless of any activists' or manufacturers' ideals or intentions, however, foam powders and tablets did little but visibly foam and cause skin irritation.

In addition to foam tablets and suppositories, which people could use without an additional barrier method, spermicidal jellies, foams, pastes, or creams could act alone but were more likely to be used with a diaphragm or cervical cap, as mentioned above. Many experimented to find the mixture of active ingredient (usually a weak acid, astringent, or solvent) and delivery base that would be most effective alongside a barrier method. After Friedrich Merz released Patentex jelly in Frankfurt-am-Main in 1911 (active ingredients were aluminum, boric acid, and chinosol), more than a hundred brands of spermicides appeared on German pharmacists' shelves by the 1920s. Spermicides became a specific field of academic research in the 1930s, with Cecil I. B. Voge's 1933 *Chemistry and Physics of Contraceptives* providing the most comprehensive guide to the efficacy and safety of the method to date. He tested 133 spermicides and authored the first study of how barrier and chemical methods interacted with each other (according to him, spermicides did not weaken rubber). Dickinson, who wrote the forward to *Chemistry and Physics*, affirmed in 1938 that he favored lactic acid and a quinine derivative in a vegetable gum base as standard ingredients for a spermicidal jelly, a formulation that

Margaret Sanger's chief physician, James Cooper, also used.[25] A breakthrough in spermicide technology came when scientists at the American pharmaceutical company Ortho patented a new spermicide in 1951.[26] The surfactant nonoxynol-9 has remained the most widely used spermicidal ingredient ever since, embedded or infused in a wide range of media.

The history of spermicides, douches, sponges, and suppositories from the mid-nineteenth through the mid-twentieth centuries illustrates several broader themes. First, there was a division between contraceptives designed and marketed for poorer women with fewer funds and infrastructural resources and middle-income and wealthier women who could maintain devices like diaphragms with running water and a temperate home. Second, the methods intended for poorer women were less effective, and they were more likely to receive poor-quality or expired goods dumped on colonies and decolonizing states from colonial, industrialized powers. Third, birth control clinics in countries with hot, humid climates faced additional problems with goods developed in cool, dry climates: the rubber on diaphragms disintegrated in hot weather and deteriorated if placed in an icebox; crows carried them off when they dried in the sun; spermicidal tablets and jellies disintegrated if poorly wrapped; suppositories absorbed dye from improper packaging; and spermicidal pastes expanded and burst out of their tubes in hot rainy weather.[27]

Contraceptive technology that fit the needs of people in diverse geographies, contexts, income levels, and intimate partnership situations did not exist.

Condoms

Although washable, reusable condoms made of animal intestines (usually the cecum of sheep or cattle), fish skins, or fish membranes were available in Europe beginning in the eighteenth century, commercial condoms quickly appeared in countries that adopted the vulcanization of rubber for medical goods in the mid-nineteenth century. Europeans gave them all sorts of nicknames: in England, they were "French letters," and in France, *capotes anglaises* (English hoods).[28] Before and after vulcanization, however, condoms had a worldwide association with prostitution and sexually transmitted disease prevention.[29] It was challenging for public health advocates to convince married couples to use them because they were associated with nonmarital sex instead of pregnancy prevention. Users also objected that their thickness dulled sensation and that putting one on interrupted spontaneity.

Though the US Comstock Act forbade the import, distribution, and sale of sex-related goods in 1873, a nationwide prohibition on widely desired items led to the expansion of an entrepreneurial black market for condoms and inspired manufacturer's creativity in marketing and advertising. The act did not forbid the importation of whole animal or fish skins or of rubber molds, so those willing to skirt the law could acquire the necessary materials for hand-rolled condoms. The most famous of these was Julius Schmid, who migrated to the United States in 1882, started an animal-skin condom business, and had his factory raided in 1890 under the Comstock Act. Despite that setback, he continued to develop new rubber

Figure 9   Ramses condoms box, Julius Schmid, Inc., New York, NY, 1929. *Source:* Courtesy Medicine and Science Collections, National Museum of American History, Smithsonian Institution, Washington, DC.

contraceptive manufacturing techniques for decades. His factory's cold-cure cement technique produced condoms that were more expensive but also more reliable than those of his competitors.[30]

Condom availability varied across Europe. Although a revision of the German criminal code in 1900 made it illegal to display or advertise "objects which are suited to obscene use," condoms, diaphragms, suppositories, and spermicidal jellies could be found widely at pharmacies and through mail order. One of the most widely used brands of German condoms was Fromm's Act, established in Berlin in 1914 by Julius Fromm. Fromm's Act condoms used Ceylon (Sri Lankan) rubber molded on glass cylinders and were dusted with a lubricant, and they could hardly be manufactured fast enough: in 1926, the company produced 24 million condoms out of 90 million nationwide. In October 1927, the German Law to Combat Venereal Diseases decriminalized prostitution and allowed advertisements for healing and preventative medical remedies as long as those advertisements were not "offensive" or "indecent." The law also allowed condom vending machines in railway stations, police stations, restaurants, cafes, nightclubs, public toilets, and military barracks. One result was that condom sales for Fromm's Act alone reached 50 million units in 1931, and references to the brand became part of piano-bar comedians' repertoires.[31]

That law was rescinded in March 1933, and Jewish business owners like Julius Fromm were forced to abandon their businesses, but access to condoms in England expanded in the 1930s. London Rubber, the company that manufactured Durex condoms, produced 2 million units per month from 1932 to 1952 using a semi-automated latex dipping system. In 1954 and 1955, the company's main factory in Chingford (eleven miles from central London) made 2 to 2.5 million per week after launching a twenty-four-hour automated system. The company focused its manufacturing on single-use, prelubricated, disposable condoms and introduced foil packaging in 1957, dominating the English market.[32]

Who bought them all? Women expected men to take care of contraceptives without much conversation if they were used at all. One woman stated that "we didn't do much talking ... he used to get them and I used to rely on him that he got them." "Aunt Polly" remembered that in Glossop, east of Manchester, in the 1930s "there was this man who came selling Durex to the men in the pub on Fridays. But we wouldn't like our husbands to use anything like that." Condoms were available by mail order; in public parks, hygiene and chemist shops, tobacconists, and barber shops; and later in the decade in hairdressers, movie theaters, and dance halls.[33] Thus, although men made most condom purchases, their availability via mail order

Figure 10   Condom vending machine, White's Comb Vendor, Inc., Elgin, IL, ca. 1955. *Source:* Courtesy Medicine and Science Collections, National Museum of American History, Smithsonian Institution, Washington, DC.

and in vending machines made them available to women as well.

World Wars I and II focused international military attention on the need for condoms in wartime. Needless to say, military regulations were intended to protect men, and not women, from sexually transmitted infections. Pairs of condoms in the form of "hygiene matchboxes" were distributed to Japanese soldiers and officers from the Russo-Japanese War in 1904 and 1905 onward. At army-authorized brothels, condom use was mandatory. From 1938 to 1945, all Japanese rubber factories were placed under military jurisdiction, and condoms were given militaristic names such as "Attack Number One" and "Attack Champion" for the Imperial Army and "Iron Cap" for the Imperial Navy. Germans also had separate field brothels for officers and rank-and-file soldiers in both world wars where condoms were required. Throughout World War II, foreign laborers were forced to produce condoms and other rubber goods in Julius Fromm's former Berlin factory, and production continued during the subsequent Soviet occupation.[34]

Condom availability in the first half of the twentieth century continued to shift worldwide due to ongoing legal, regulatory, military, and technological changes. Although the Comstock Act stayed on the US books until 1937, its power was weakening due to further judicial rulings and limited enforcement. A major victory for American birth

Military regulations
were intended to protect
men, and not women,
from sexually
transmitted infections.

control advocates came in 1918, when a New York federal appeals court judge ruled that physicians could prescribe contraception to prevent disease (but not to prevent pregnancy). Moreover, the discovery of noncombustible latex in 1920 and the uniform condom ring machine in 1926 (which eliminated the need for hand rolling) shifted the technological landscape of barrier contraceptives once again and made it possible for manufacturers to make sturdier condoms more quickly and safely. In 1938, a revised Federal Food, Drug, and Cosmetic Act placed the condom, as a disease preventative, under the jurisdiction of the US Food and Drug Administration. Only two condom firms could afford the manufacturing changes necessary under the increased federal quality control standards, and others either closed or shipped their lower-quality goods overseas.[35] So although condoms were available in many industrial and decolonizing countries by the time the hormonal pill would appear, access, quality, cost, and use varied significantly. The dual function of condoms as both pregnancy and disease preventatives gave their use an ambiguous status and meaning.[36] Researchers continued to investigate other options, including the more invasive IUD.

IUDs

The intrauterine device, or IUD, was designed to block sperm or to create a hostile environment within the uterus so that an embryo could not implant. Such a method, based on the nineteenth-century stem pessary that was used for uterine problems, would have advantages for some women: it did not need to be cleaned and stored after each use; it did not require a husband's permission or consent; and it did not require preplanning. The first American patent for an IUD in that shape was issued to a George J. Gladman of Syracuse, New York, in 1895, but it is unclear if he ever manufactured it.[37] German doctors, including a Dr. Hollweg in Magdeburg, experimented with IUD designs fashioned with stem pessaries beginning in 1903, though he was criminally charged with negligence causing bodily harm to patients for doing so. A Dr. Richter in Waldenburg also experimented with these devices beginning in 1909, the same year as Ernst Gräfenberg. Gräfenberg decided to abandon the stem pessary form altogether and fashioned a new kind of IUD with silkworm gut and silver wire laced with copper.[38] After first hearing about the "Gräfenberg ring," London-based physician Norman Haire advocated it instead of the diaphragm and jelly method, even though 13 percent of them extruded.[39] He tried to use his support of the Gräfenberg ring as a means to gain funding for further research on the device from the Birth

Control Investigation Committee in 1927. The BCIC gave the money instead to Helena Wright, who attempted but later abandoned an attempt to manufacture her own IUD with coils covered in India rubber.[40] Gräfenberg publicized his ring at European sexological congresses in 1929 and 1930, and he continued to tinker with it in the early 1940s after moving to the United States.[41]

Even before the IUD got the attention of birth controllers around the world at these congresses, the disadvantages of the device were evident. The contraceptive advocate Ettie Rout, who gathered observations about different methods in her 1922 book *Practical Birth Control*, pointed out that the IUD caused sepsis, miscarriage, stillbirth, and pain and could expulse spontaneously from the uterus. Dickinson's contraceptive guidebook included images of multiple IUD shapes and the different ways they could harm women, such as increased bleeding, heavy menstruation, and uterine perforations. The Danish physician and birth control advocate Jonathan H. Leunbach experimented with silver IUD rings that caused patients only bleeding and pain.[42] A gold-plated IUD designed by the Japanese physician Ōta Tenrei in 1932, in addition to being painful and ineffective, also caused infections and infertility.[43] Some physicians continued to experiment with and to prescribe IUDs in the 1940s and 1950s, but there was little innovation in their design, materiality, and manufacture until the 1960s. Their disadvantages far outweighed their advantages, and many women instead chose to manage their fertility without professional medical assistance.

Herbs

Women across the world have used herbs in teas or douched with them to bring on menstruation and to avoid pregnancy. Herbals, or texts with recipes for homemade medicines, were popular throughout Europe and colonial and early America from the seventeenth through the first decades of the twentieth century. They recommended preparations called *emmenagogues* with pennyroyal, rue, savin, tansy, and ergot of rye to combat amenorrhea, or a lack of menstrual flow, which could either prevent pregnancy or could end a very early one before fetal movement began.[44] A doctor from the German town of Neuenahr (now Bad Neuenahr-Ahrweiler) recorded the words of children's songs in Bavaria, the lower Rhine Valley, and Brandenberg (near Berlin) in the late nineteenth century. They sang of the herbal wreaths that women wore in their hair during a wedding and could also drink as a tea to prevent pregnancy: "Rosemary and thyme, grows in our garden, young Aennchen is the bride and can wait no longer. Red wine and white wine, tomorrow is the wedding." Herbal variations in the first line included lavender, myrtle, parsley, and chervil. Another song he heard was a short dialogue between a woman and a gardener: "Good day Mr. Gardner, do you have lavender, rosemary and thyme, and a little wild thyme? Yes, Madame! We have all of these, outside in the garden."[45] Whether or not German women actually used these herbs for these purposes is hard to know, but these songs preserved an association between women and herbs used as emmenagogues.

Herbs used to bring on or to suppress menstruation varied across the world. Ergot of rye was available in northern and eastern Germany because rye was widely grown for bread. Pennyroyal and rue, along with aloes, wild celery, and bracken fern, were also known among European colonists in South Africa, and Welsh and English people who were sexually active between 1925 and 1950 joked and gossiped about using slippery elm and pennyroyal to end unwanted pregnancies. From the mid-nineteenth to the early twentieth centuries in South Africa, Malaysian immigrant women used preparations with red geranium, Khoisan women used a type of thornbush, and Zulu women used a peppery shrub called *uhlungughlungu*. A 1952 Planned Parenthood Foundation of America report found that nurses and working-class women in Jamaica boiled herbs such as ram goat rose, pennyroyal, pepper elder, or rice bitter, sometimes with rusty nails, in a tea to drink. These methods likely originated during the slavery era in the Caribbean.[46]

Although knowledge of pennyroyal has appeared on and off in the historical record for almost a millennium, it is hard to know its meaning to the women who used it. One clue appears in Sarah Orne Jewett's 1896 novel *The Country of Pointed Firs*, which centers on the experiences of a summer boarder in rural Maine. In the chapter "Where the Pennyroyal Grew," the locally respected herbalist and widow Almira Todd invites the unnamed narrator to visit a small island where her mother lives. On a walk, they come across a patch of pennyroyal, which delights Mrs. Todd and triggers her memory:

> My heart was gone out o' my keepin' before I ever saw Nathan; but he loved me well, and he made me real happy, and he died before he ever knew what he'd had to know if we'd lived long together. ... I always liked Nathan, and he never knew. But this pennyr'yal always reminded me, as I'd sit and gather it and hear him talkin'—it always would remind me of—the other one.[47]

Mrs. Todd's language is veiled, but she may be referring to using pennyroyal to end a pregnancy with a lover who preceded her husband. Jewett's father was a physician, and Jewett herself was a gardener and familiar with medicines that rural Mainers used—including emmenagogues and abortifacients. The pennyroyal scene highlights tensions in medicine at the turn of the last century: on the one hand, women could manage their fertility if they had the correct knowledge; on the other hand, using this knowledge could have damaging or deadly results if used improperly.[48]

At the same time that *Country of Pointed Firs* was published, pharmaceutical corporations in Western countries were adapting ingredients from homemade medications into marketable, physician-controlled therapeutic products. In the United States, the pharmaceutical company Parke, Davis advertised liquid and pill emmenagogues in its catalogs using traditional ingredients such as cotton root bark, ergot, pennyroyal, rue, and tansy from 1898 through 1937. They also contained a hefty dose of alcohol—up to 75 percent. In Germany, the Prussian Police Ordinance on Trade with Poisonous Substances, which came into effect in 1894 and was revised in 1904 and 1906, limited the sale of "poisonous substances," which often included the herbal remedies formerly made at home.[49] The use of herbal methods continued largely unrecorded throughout the twentieth century, though records of women using them waned.

Timing Methods

Even less expensive than herbal preparations, contraceptive timing methods have been in use for centuries. For

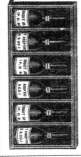
Figure 11 Standard Fluid Extract Ergot advertisement and Ergot Aseptic advertisement in a Parke, Davis & Co. catalog, 1898. *Source:* Parke, Davis Series, Trade Literature Collection, National Museum of American History Library, Smithsonian Institution, Washington, DC.

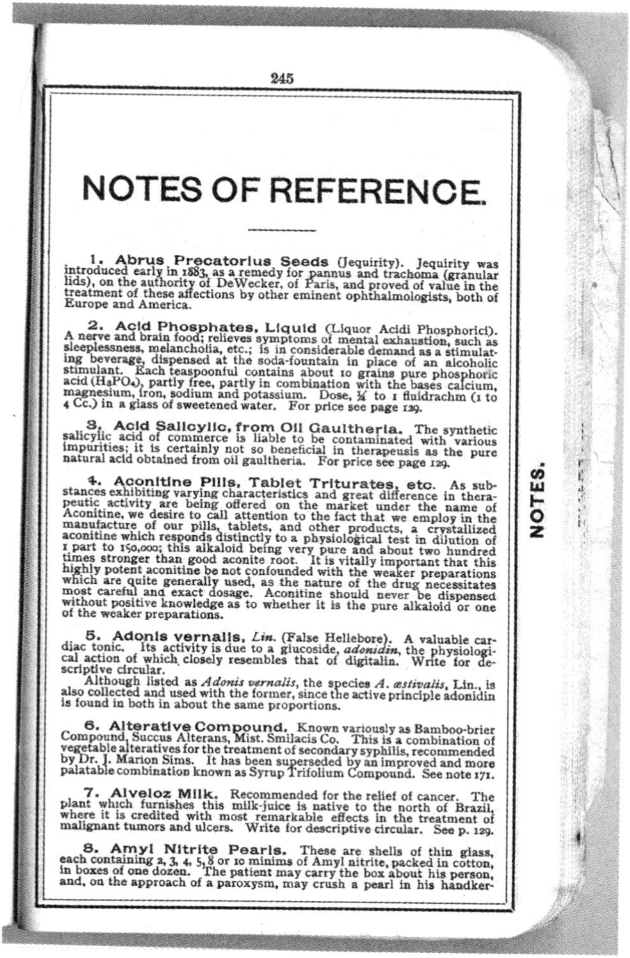

NOTES OF REFERENCE.

1. **Abrus Precatorius Seeds** (Jequirity). Jequirity was introduced early in 1883, as a remedy for pannus and trachoma (granular lids), on the authority of DeWecker, of Paris, and proved of value in the treatment of these affections by other eminent ophthalmologists, both of Europe and America.

2. **Acid Phosphates, Liquid** (Liquor Acidi Phosphorici). A nerve and brain food; relieves symptoms of mental exhaustion, such as sleeplessness, melancholia, etc.; is in considerable demand as a stimulating beverage, dispensed at the soda-fountain in place of an alcoholic stimulant. Each teaspoonful contains about 10 grains pure phosphoric acid (H_3PO_4), partly free, partly in combination with the bases calcium, magnesium, iron, sodium and potassium. Dose, ¼ to 1 fluidrachm (1 to 4 Cc.) in a glass of sweetened water. For price see page 129.

3. **Acid Salicylic, from Oil Gaultheria.** The synthetic salicylic acid of commerce is liable to be contaminated with various impurities; it is certainly not so beneficial in therapeusis as the pure natural acid obtained from oil gaultheria. For price see page 129.

4. **Aconitine Pills, Tablet Triturates, etc.** As substances exhibiting varying characteristics and great difference in therapeutic activity are being offered on the market under the name of Aconitine, we desire to call attention to the fact that we employ in the manufacture of our pills, tablets, and other products, a crystallized aconitine which responds distinctly to a physiological test in dilution of 1 part to 150,000; this alkaloid being very pure and about two hundred times stronger than good aconite root. It is vitally important that this highly potent aconitine be not confounded with the weaker preparations which are quite generally used, as the nature of the drug necessitates most careful and exact dosage. Aconitine should never be dispensed without positive knowledge as to whether it is the pure alkaloid or one of the weaker preparations.

5. **Adonis vernalis,** *Lin.* (False Hellebore). A valuable cardiac tonic. Its activity is due to a glucoside, *adonidin,* the physiological action of which closely resembles that of digitalin. Write for descriptive circular.

Although listed as *Adonis vernalis,* the species *A. æstivalis,* Lin., is also collected and used with the former, since the active principle adonidin is found in both in about the same proportions.

6. **Alterative Compound.** Known variously as Bamboo-brier Compound, Succus Alterans, Mist. Smilacis Co. This is a combination of vegetable alteratives for the treatment of secondary syphilis, recommended by Dr. J. Marion Sims. It has been superseded by an improved and more palatable combination known as Syrup Trifolium Compound. See note 171.

7. **Alveloz Milk.** Recommended for the relief of cancer. The plant which furnishes this milk-juice is native to the north of Brazil, where it is credited with most remarkable effects in the treatment of malignant tumors and ulcers. Write for descriptive circular. See p. 129.

8. **Amyl Nitrite Pearls.** These are shells of thin glass, each containing 2, 3, 4, 5, 8 or 10 minims of Amyl nitrite, packed in cotton, in boxes of one dozen. The patient may carry the box about his person, and, on the approach of a paroxysm, may crush a pearl in his handker-

Figure 11 (continued). Ergot preparations were advertised alongside those for cancer, secondary syphilis, and other ailments.

example, Henry A. Allbutt's 1887 pamphlet *Wife's Hand-book: Pregnancy and after Delivery* recommended temporary abstinence five days before and eight days after a menstrual period.[50] Members of some cultures believed that prolonged breastfeeding curtailed the ability to become pregnant, though it is more likely that taboos about sex with nursing mothers were the real reason for lower numbers of pregnancies while breastfeeding.[51] However, before the Japanese gynecologist Ogino Kyūsaku and his Austrian counterpart Hermann Knaus pinpointed the time of ovulation in 1924 and 1929, respectively, any timing method beyond complete abstinence was guesswork.[52] Although timing methods required no money and no physical technologies, they demanded control and a willingness to deny oneself pleasure that not all were able to manage.

Coitus reservatus and alternative sexual positions were thought to minimize the possibility of sperm and egg connecting in penile-vaginal intercourse. *Coitus reservatus* involved male tumescence and detumescence in the vagina without ejaculation for up to an hour, and the mid-nineteenth century Oneida utopian community in upstate New York used the practice to limit births. The Chicago-based physician Alice B. Stockham made more Americans aware of the practice a few decades later with her books *Tokology* (1893) and *Karezza* (1896). Oral sex and interfemoral intercourse were also possibilities, and some marriage manuals provided detailed instructions about positioning. Nineteenth-century Germans used a method called *coitus obstructus* or *coitus saxonicus*, which meant pressing the base of the penis at the moment of ejaculation in order to force semen into the bladder. An Estonian marriage manual called *How to Avoid Pregnancy* (*Kuidas Hoiduda Rasedusest*) published in 1934 recommended that heterosexual couples have sex on their sides and that when the man was ready to ejaculate, he should pull partway out and both partners should spread their legs. Thus, the ejaculate would land closer to the outside of the vagina and could be washed away easily. Marie Stopes recommended that women sit up immediately after men's ejaculation and cough "violently" to contract their pelvic muscles in an emergency if no other methods were available, and middle- and upper-income Indian women were advised to do jumping jacks after coitus to expel semen. As some believed that women's orgasm was the mechanism for semen to move into the uterus, women in the United Kingdom, the United States, and Czechoslovakia heard or read erroneous advice to avoid orgasm as a means of avoiding pregnancy.[53]

Sex between two men and two women, masturbation, and anal coitus were other options altogether, but English-language contraceptive advocates disapproved of them when they mentioned them at all. Dickinson believed that primarily "laborer and peasant husbands" used anal coitus, that oral sex was the province of prostitutes, and that vulvar and interfemoral coitus were only premarital practices. "All of the above fall into the class of contraceptive measures to be deplored or displaced," he sniffed. Ettie Rout claimed that homosexual sex and masturbation were "anti-social ways of controlling fertility." In Germany, however, doctors in the 1890s tracked the rising prevalence of anal gonorrhea among women.[54] It is impossible to know if those women were engaging in anal sex as a form of contraception, for pleasure, or both, but either way, doing so had the unfortunate result of a sexually transmitted infection.

Yet another timing method, withdrawal, required trust in men's self-control and judgment. The historian Norman Himes asserted that *coitus interruptus* "is doubtless the most popular, widely diffused method of contraception." In mid-nineteenth century Søgne, Norway, the practice was called *hoppe av i farten* (jump off while the going is good), and women in mid-twentieth-century Trinidad, Barbados, and Jamaica referred to it in birth control clinic interviews as "my husband is careful." In early twentieth-century southern Wales, it was men's responsibility to decide on the timing of sexual activity and contraception; to practice withdrawal, one would "take the kettle off the fire before it boils over."[55] In Meru, Kenya, a native-born Methodist minister reported in the 1920s and 1930s that young people had *coitus interruptus* in a standing position.

In mid-nineteenth century Søgne, Norway, the practice [of withdrawal] was called *hoppe av i farten* (jump off while the going is good).

A survey of 11,126 Czechoslovakian women in 1956, most of whom were between twenty and thirty-nine years old, found that 68.4 percent of them used withdrawal as their primary contraceptive method. Stopes opposed this method based on her belief that women received nutrients from ejaculation, and thus "the woman subject to this process is also deprived of the possibility, after the union is completed, of the beneficial absorption from the seminal and prostatic fluid."[56]

The Roman Catholic Church disliked the method for yet another reason. According to church law, penile-vaginal coitus, or "the marital act," was permitted only between a man and a woman married to each other for procreation. A marital act thus required male ejaculation, so the near-universal method of withdrawal was nonetheless sinful. Since the mid-nineteenth century, however, church leadership had quietly permitted married couples to abstain from sex completely from time to time in order to space their children, largely as a means of avoiding withdrawal.[57] Any artificial contraceptives, such as condoms or diaphragms, "frustrate the marital act" and so were also sinful. Pope Pius XI affirmed this perspective in the 1930 encyclical *Casti connubii* (*On Christian Marriage*), and his successor, Pius XII, officially approved the rhythm method based on women's menstrual cycles in an October 1951 address to Italian midwives.[58]

Complete abstinence was yet another option, but evidence of its use is rare and anecdotal. The Finnish author Arvid Järnefelt wrote in his memoir *Vanhempieni romaani* (*My Parent's Story, 1928–1930*) that after his mother gave birth to nine children, his father moved his bed out of their shared bedroom to signal that their sexual interactions had ended for good. A woman interviewed in southeast Norway recalled that a doctor advised her parents to abstain from sex after her mother was seriously weakened from giving birth to five children in seven years (1903 to 1910). Her father then moved to South America, where he remained for ten years.[59] For many, the disadvantages of abstinence clearly outweighed the advantages.

Abstinence took on a different cast in India, where it was a central element of *brahmacharya*—the idea that one must control lustful desires. In the early twentieth century, Mahatma Gandhi advanced the idea that overindulgence of the body was a specific curse of modern Western civilization, and barrier or behavioral contraceptives did nothing to tame the real problem: an overabundance of sexual desire, particularly male desire. Women, in turn, should refuse their husband's advances. Any child after the first was born of lust, not duty or religion (dharma), and procreation should be limited until India was free from colonial rule. He met Margaret Sanger during her November and December 1935 visit to India, and she thought that she convinced him that the safe period was acceptable.

The Indian sexologists A. P. Pillay and N. S. Phadke thought that Gandhi's beliefs were acceptable (if they did not cause harm) and ridiculous, respectively.[60] After partition in 1947, Indians and Western birth control promoters alike turned to advocating sponges, spermicides, and diaphragms instead of timing methods and celibacy.

Sterilization

Although crudely cutting off or maiming the genitals of criminals was not unknown before the twentieth century, sterilization in the early twentieth century became linked with eugenic approaches to controlling populations. Those whom physicians, government officials, prison wardens, and other authority figures deemed "unfit" to reproduce— including un- or undereducated people, recent migrants, criminals, poor whites in rural parts of the US, physically or mentally disabled people, convicted criminals, and people of color—were subject to sterilization beginning in the 1900s. Men and women alike could undergo sterilization, sometimes without their knowledge or consent, for a range of reasons. Sterilization of male genitalia was a relatively simple and short operation: a physician severed the vas deferens, which eliminated the sperm after two weeks and could be reversed later. Sterilization of female genitalia, harder if not impossible to reverse, required major abdominal surgery, either an ovariectomy (the excision of ovaries) or a salpingectomy (an abdominal incision and tying or severing the fallopian tubes).[61]

The practice of sterilization in prisons began with the coincidentally named Harry C. Sharp, medical superintendent of the Indiana Reformatory in Jeffersonville. He first severed the vas deferens of a male prisoner who complained of uncontrollable masturbation in 1899; he then sterilized 450 prisoners before successfully lobbying the state legislature to pass the first US law mandating the sterilization of the "unfit" in 1907. Twenty-six states followed Indiana's example, and the right of states to coerce sterilizations was upheld in the now-notorious case *Buck v. Bell*, 274 US 200 (1927). The majority opinion in the case, involving the forced sterilization of seventeen-year-old rape victim Carrie Bell in Lynchburg, Virginia, upheld the state's right to sterilize "imbeciles" in the name of protecting public health. Following the *Buck v. Bell* decision, the scope of the Indiana law expanded, and minors as young as sixteen in state institutions were targeted for sterilization. Between 1907 and 1937, 27,869 Americans were sterilized: 16,241 men and 11,628 women.[62]

The practice was not confined to the United States. Vasectomy was introduced into Germany in 1894, and a method of salpingectomy was introduced in 1910. Doctors could perform them for "therapeutic purposes," and they became a favorite method of eugenicists before

World War I. The New Zealander Ettie Rout stated a standard elite perspective matter-of-factly: "suitable cases for sterilization are mental instability, hereditary taints, tuberculosis, syphilis, [and] repeated and overfrequent pregnancies which are undermining the sound health and economics of the home." The physician Norman Haire tried to sterilize women hormonally, but when that did not work, he began to advocate publicly in the 1920s for the "sterilization of the unfit in the interests of the race." He also tried to market vasectomies in the 1930s as "male rejuvenations." Also between 1931 and 1934, the British Eugenic Society supported a coerced sterilization bill, but it lacked parliamentary support. Most notoriously, during the Nuremberg Trials, Nazis on trial for war crimes cited *Buck v. Bell* as part of their reasoning for sterilizing two million Germans during the Third Reich.[63]

Sterilization had a well-deserved reputation in Europe and beyond as a coercive eugenic operation. In the Nordic countries—Denmark, Norway, Sweden, and Finland—an increase in involuntary or coerced sterilization correlated with the rise of the welfare state. Limiting the ability of mentally disabled people to reproduce began piecemeal at Nordic state institutions in the 1910s but became national law in the 1920s and 1930s. Both adults and minors could be sterilized if they had low mental abilities, could not support potential children financially in the future, or presented a risk of passing on a hereditary disease to offspring.[64] Rather than decreasing after World War II, sterilization across the region increased, especially in Sweden and Finland, where it could be carried out on eugenic, social, or general medical grounds. Working mothers with large families who exhibited "weakness" or who were in "social distress" in both countries were specifically targeted, and 99 percent of the 56,080 sterilizations in Finland were carried out on women. Health authorities thought they were less likely than men to refuse consent.[65]

Forced sterilization laws in the United States and the Nordic countries stayed in place until patients' rights activism began in the 1960s and 1970s and abortion was legalized (chapter 4). The persistence of laws and practices in democracies that discriminated against people with disabilities would later highlight the need for global affirmation of every individual's right to have children (chapter 5). Democratic governments alone did not protect the most vulnerable from coerced and involuntary sterilization. It would take declarations of human rights on a global stage and citizen organizing to challenge the practice.

Conclusion

This overview of contraceptive technology before the hormonal pill illustrates the range of methods that people have used to prevent and to space pregnancy. Contraceptive use was intimate but shaped by political, technological, religious, and sociocultural forces. The use of a technique or technology depended significantly on religion, pain tolerance, reversibility, imprisonment, mental health, access to technology and professional care, knowledge of methods, income, correct use, effectiveness, personal motivation, and relationship status, among other reasons. Contraceptive technology could be temporary or permanent, forced or voluntary, painful or pleasurable, effective or pointless. The ability to control and to time when a pregnancy occurred was of such importance that the introduction of a new pharmaceutical contraceptive, the hormonal pill, would have worldwide repercussions that continue to the present.

THE PILL AND ITS SUCCESSORS

The development of hormonal contraception forever changed human dynamics related to reproduction. First available only as a pill, hormonal contraception was soon available in a wide range of doses and forms, such as an implant, an intramuscular shot, a patch, a ring, and as emergency contraception after unprotected sex. Its creation, manufacture, sale, and dissemination also changed the perspective and work of family planners and population controllers, along with relationships between citizens and their governments, between doctors and patients, and between long- or short-term sexual partners. It also shifted how people thought about their bodies, sexualities, erotic relationships, and reproduction.

This chapter describes the discovery of the hormonal method of contraception, its rapid spread around the world, its effects on health, and the ways it altered human

The development of hormonal contraception forever changed human dynamics related to reproduction.

relations related to sex and reproduction according to the cultural contexts in which people adopted and used it. Its most arresting feature was that it was now possible, "for the first time ever, to separate contraception from the act of sexual intercourse."[1] This chapter highlights the bumpy acceptance of hormonal methods in three different contexts: the United States, Communist countries during the Cold War, and two countries in sub-Saharan Africa—Zimbabwe and South Africa. It concludes with how different formulations and delivery methods of contraceptive hormones—such as the Depo-Provera shot, emergency contraception, and the male version of the pill—had specific effects on human health, behavior, and thought beyond the medicine's first wave.

Founding Hormonal Contraception

The discovery of hormonal contraception has been described in extensive detail.[2] Though there was wide interest in the early twentieth century in finding a safe, effective, stable, and inexpensive form of contraception that a wide range of people could use, the major American and global agencies that invested in medical research in the 1940s and 1950s—such as the US National Institutes of Health (NIH), the US National Science Foundation (NSF), and the World Health Organization (WHO)—did not support contraceptive research due to its association with sex. The energy for contraceptive research (as opposed to the distribution of and tinkering with existing methods like spermicides and diaphragms) came not from nonprofit, nongovernmental organizations but from scientists working for pharmaceutical companies. They were looking for ways to extend the discovery of synthetic steroids for treating a range of maladies in new directions. Two of the companies that funded scientists were Syntex and G. D. Searle, and the Austrian scientist Carl Djerassi, who synthesized cortisone from the roots of a Mexican yam, was among the scientists pursuing a breakthrough.

The individuals involved in the discovery of the hormonal pill each contributed something critical to the process. Katharine McCormick, the first female graduate of the Massachusetts Institute of Technology with a bachelor's degree in biology, was married to Stanley A. McCormick, the heir to the International Harvester Company. In the 1910s, Margaret Sanger asked McCormick to help with early contraceptive advocacy efforts by buying diaphragms in Europe and smuggling them back to the US, which she likely did.[3] Following Stanley's death in 1947, Katharine devoted herself fully to women's causes, including donating part of her fortune to contraceptive research. She contacted Sanger in 1950 about the best ways to support it, and Sanger put her in touch with Gregory Pincus.

Pincus founded the independent Worcester Foundation for Experimental Biology in 1944 after he was denied tenure at Harvard University. In addition to McCormick's support (which eventually totaled approximately $2 million), he also had a grant from Searle, who hired him to find an efficient, inexpensive way to synthesize the hormone cortisone for arthritis.[4] Pincus began research on hormones in 1951 with a colleague, Min-Chueh Chang, finding that progesterone—synthesized from the same Mexican yam root that Djerassi used for cortisone—halts contraception by preventing the release of hormones needed for egg maturation.

Djerassi and a chemist at Searle both made progesterone into a pill form, and Pincus, along with the obstetrician/gynecologist John Rock, began testing Searle's version of the pill, called Enovid. The testing took place with some of Rock's patients, involuntary psychiatric patients at Worcester State Hospital in 1954, and then with volunteer nurses at Puerto Rico hospitals in April 1956. The two female physicians involved in the Puerto Rican testing, Edris Rice-Wray and Penny Satterthwaite, drew attention to the severe side effects (headaches, nausea, cervical erosion, breakthrough bleeding, and dizziness) that caused many nurses to drop out of the trial, but Pincus and his male collaborators downplayed them and supported Searle's bid to the US Food and Drug Administration (FDA) to approve the drug.[5] The FDA approved

Enovid (9.85 milligrams of the progestin norethynodrel and 150 micrograms of estrogen) in June 1957 for gynecological disorders; however, doctors who were willing to prescribe the pill off-label as a contraceptive could already begin to do so. The FDA then approved the pill for contraception in June 1960 and limited use initially to two years.[6] Once that wider-use authorization occurred, the manufacturing, importation, and sale of the pill spread rapidly.

Tracing the Impact of Hormonal Contraception

Although hormonal contraception had worldwide effects, it manifested differently in diverse areas depending on the individuals, cultures, religions, and governments involved in regulating, distributing, and monitoring its use. The pill's effects in three areas are traced below: the city of Lawrence, Kansas, followed by protests at the US federal level; Communist countries during the Cold War; and two sub-Saharan African countries—Zimbabwe and South Africa. The use of an alternative hormonal delivery method, the Depo-Provera shot, had clear reverberations across apartheid-era South Africa. Together, these snapshots of hormonal contraception's reception show the complexity of introducing a new technology that had the potential to upend so many individual, doctor-patient, societal, and

Figure 12 Ortho-Novum Pharmaceutical "Dialpak" contraceptive pill dispenser, Raritan, New Jersey, 1963. *Source:* Courtesy Medicine and Science Collections, National Museum of American History, Smithsonian Institution, Washington, DC.

government-citizen relationships. It changed pill-takers' relationships to their own bodies, their partners, and their medical care, as well. Hormonal contraception was also a new kind of medicine—one that had to be taken regularly and could potentially be taken for decades by healthy women. Scientists, physicians, family planners, population controllers, pharmaceutical companies, and women taking the pill all participated in a scientific experiment that was at once wide-ranging and international as well as deeply personal.

The pill had a mixed impact on the lives and behaviors of Americans in towns like Lawrence, Kansas, which was home to the University of Kansas and approximately 33,000 people in 1960. After the pill was approved for contraceptive use that June, it was soon available in private clinics and at the university health center only to married students. The director of the Lawrence health department offered the pill without requiring a pelvic exam to married, non-collegiate women in June 1965, as soon as it was legal to do so—a decision that only 20 percent of health departments nationwide made. The director's personal motivation was to establish a liberal distribution policy in order to assist with the global problem of population control.[7] In the local press, however, his decision was akin to the public acceptance of premarital sex, thus disrupting traditional rules and expectations of religion and morality. Local women appreciated the liberality of health department policy but questioned the risks of distributing pills without a pelvic exam to check if there were any reproductive health conditions that might preclude pill use.

Later, in February 1972, twenty women with their children highlighted uneven contraceptive availability across Lawrence by staging a sit-in at the university health center, demanding that it prescribe the pill to unmarried female students and offer pap smears and gynecological exams.[8] The health-care center changed its policy for students after the sit-in, and for nonstudents, the Lawrence health department became a fully functioning family planning clinic in 1974 after a change of leadership. Thus, obtaining the pill depended on whether a pill seeker attended the university, whether she was married, whether she had a private physician or used public services, and whether she (if unmarried) felt comfortable taking the pill in the face of public disapproval for premarital sex. From examining the situation in Lawrence, it is clear that the pill did not change history all at once but instead that change took place unevenly across the country depending on individual beliefs and sociopolitical contexts.

The FDA approval of the pill also had broader effects on medications prescribed in the United States. In 1969, the journalist Barbara Seaman published *The Doctor's Case against the Pill*, which was an exposé of the pill's side effects and doctors' failure to inform women of them.[9] Seaman detailed how women taking the pill developed diabetes, blindness, severe depression, hypertension, and strokes and sometimes even died. She pointed out that women did not have access to unfiltered information about the pill's side effects, doctors often did not tell patients about those side effects, and drug companies provided doctors with a profit motive for prescribing the pill instead of diaphragms or condoms. The Wisconsin senator Gaylord Nelson, chair of the Senate Committee on Small Business Practices, read Seaman's book and decided to hold hearings on the pill with only male witnesses in January 1970. Feminist protesters from the D.C. Women's Liberation group disrupted the hearings, demanding that women also be permitted to testify and holding a spontaneous news conference afterward.[10] There were two significant end results of the hearings and protests: all medications distributed in the US thereafter were required to have a printed insert listing the proper dosage and side effects, and the American feminist women's health movement began to organize to make lasting changes to a medical establishment that they perceived as chauvinist and patriarchal.[11]

The pill had a different reception in Eastern Bloc countries. Each Communist government's interest in promoting or limiting births for specific ethnic groups determined the availability of contraception beyond abstinence and withdrawal.[12] A country's ability to provide the pill for its citizens also depended on its ability to produce it in sufficient quantities because imports from Western countries were expensive. Both the pill and IUD appeared in Eastern Europe in the second half of the 1960s: in the former East Germany in 1965, in the former Czechoslovakia in 1966, in Hungary in 1967, and in the former Yugoslavia in 1968. Imported pills were available in Poland in the early 1960s, and state-sponsored manufacturing began in 1969.[13] Citizens of Communist countries also had varying opinions about whether to use the pill, depending on cost and accessibility generally and their broader view of the West. Although the East German government provided the pill for free to citizens, thus signaling state approval of the method, there were public fears about the pill's long-term side effects that countered official support.

Of all Communist governments, Russia's was most opposed to the pill. The national Ministry of Health sent letters cautioning citizens against using the pill in 1974 and again in 1981, greatly exaggerating the potential health problems accompanying pill use. Abortion remained the most widely used method of preventing a pregnancy from developing through the end of the Cold War in Russia.[14] Across the Atlantic in Cuba, only a small number of women used the pill. They were dependent on Russian imports for pills in the 1970s and 1980s, and the US embargo made it challenging to depend on a steady, affordable supply. The dependence on Russian imports created two specific obstacles to widespread pill adoption in

Cuba. First, the Soviet imports were manufactured with the higher dosages of estrogen and progesterone first developed in the 1950s instead of the lower-dose formulations found to be effective later. So, Cuban women taking the pill were more likely to suffer side effects. Second, because the supply of pills was undependable, Cuban women who had access to the pill would cycle on and off of it due to its inconsistent availability, and they experienced health problems as a result.[15]

Even though the pill became more obtainable and less expensive throughout the 1970s and 1980s, many Eastern European couples continued to practice withdrawal. The pill came on the Czech market in 1966, but by 1977 only 5 percent of Czech women were using it. A 1986 survey of three thousand women in Serbia indicated that only 22 percent used the pill or an IUD.[16] Many people perceived the pill as a symbol of Western excess and decadence and as a source of Western-influenced changes in gender dynamics because it allowed women to subvert their male partner's desire to have children without his knowledge or consent. Furthermore, some men saw the effective use of withdrawal as a symbol of their prowess and ability to control their behavior during intercourse. If a pregnancy happened, it was then the woman's problem to manage, which she often did with an abortion.[17] These conditions—state support of abortion facilities, the limited availability of barrier or hormonal contraceptive

Many people perceived
the pill as a symbol
of Western excess
and decadence.

technologies (particularly outside cities), and traditional gender dynamics in marriages—all produced a climate in which many women living behind the Iron Curtain managed their fertility with abortion instead of contraception.

The advent of hormonal contraception also affected racial, gender, and family power relations in other countries. For example, in colonial Rhodesia (now Zimbabwe), Shona women managed their fertility by wearing a belt with knots tied in it to represent the number of years that they aimed to be childfree. A woman's husband would untie the belt to encourage fertility again. A husband would also use withdrawal to avoid conception, and healers would give both partners herbs to drink or to tie around their waists if the healer thought that another pregnancy would harm the woman. Furthermore, a woman was responsible for producing healthy heirs for her husband and his family, so her fertility was the business not only of the marital partners but also of her in-laws. Over several years in the 1950s, however, "the gatekeepers of the old methods—husbands, mothers-in-law, and traditional healers—were replaced by a new set of gatekeepers for the new methods—clinic staff, community-based distributors, and nurses."[18]

These new gatekeepers included the minority white Family Planning Association (FPA), which began in the 1950s to distribute spermicidal Volpar foaming tablets and information about their use among the native Shona population. The foaming tablets were much more popular than IUDs or diaphragms. White women could be prescribed the hormonal pill under the brand Anovlar in May 1961, but it did not become available for Shona women until a few years later. The FPA made the Depo-Provera shot available to those visiting its clinics in the early 1970s, and it became more popular than the pill in 1974. Over a period of two decades, "the new [Western] contraceptives represented a reallocation of power over African childbearing, away from African families and communities, and toward the white colonial state."[19] After Zimbabwe became independent in April 1980, the gender and family power relationships that were shifting as a result of women using hormonal contraception became part of the struggle over the direction of postcolonial Zimbabwe. As one sociologist writes, "in addition to removing power over childbearing from men, the pill and the injection unleashed many other anxieties among men, such as fears of dangers of unleashed female sexuality."[20] A 2006 to 2011 study showed that Zimbabwean women's chosen technology differed greatly according to marital status: married women largely used oral contraceptives, and never-married women mostly used condoms.[21] The condom's dual advantage as a device to prevent pregnancies and sexually transmitted infections (STIs), along with its wide availability and lack of side effects, make it a more appealing option to many.

Hormones in a Different Form: The Depo-Provera Shot

The approval of hormonal contraception in the United States in June 1960 spurred scientists and pharmaceutical companies around the world to get their own formulations of progesterone and estrogen onto the market as quickly as possible. The side effects that Rice-Wray and Satterthwaite noticed in the Puerto Rican pill trials had not disappeared, however, and scientists began to tinker with lower pill dosages to minimize side effects and risk. The original dose of Enovid was 9.85 milligrams of progestin and 150 micrograms of estrogen, and by 2012, the average dose was 0.1 to 3.0 milligrams of progestin and 20 to 50 micrograms of estrogen.[22] Throughout the twentieth century, other scientists experimented with alternate forms of delivery that would not require women to take a pill at the same time every day—with varying degrees of success.[23]

One of those delivery methods was a hormone shot. In the process of developing steroids for pain relief, the company Upjohn created a shot of medroxyprogesterone acetate (MPA) that a user would take every ninety days to maintain a steady supply of progestin. The FDA approved it in the US to treat endometriosis and miscarriage and granted permission to Upjohn to test it as a contraceptive beginning in 1963. Depo-Provera had similar side effects as those found in high-dose contraceptive pills, including blood clots, depression, menstrual irregularities, and weight gain. The FDA granted and then withdrew approval for Upjohn's new drug application for limited marketing in 1974, though the company continued to pursue that approval throughout the 1970s and 1980s. In the meantime, major family planning agencies, including the WHO, the United States Agency for International Development (USAID), and the International Planned Parenthood Federation (IPPF) supported Depo-Provera and distributed it in eighty other countries.[24]

The American feminist health movement, organized as the National Women's Health Network (NWHN) in 1975, mobilized against Depo-Provera. That mobilization happened alongside widespread reporting about the damage that the Dalkon Shield caused (chapter 4) and reports that DES (diethylstilbestrol), a synthetic estrogen, caused a usually rare vaginal cancer in the female children of women who were pregnant while taking it to prevent miscarriage. Throughout the 1970s and early 1980s, NWHN established a registry of Depo-Provera users who would testify about how the shot harmed them and held regular press conferences to organize public opposition. When Depo-Provera was brought before the FDA's Public Board of Inquiry in January 1983, the NWHN provided opposition testimony.[25] In part due to the NWHN's actions, the FDA did not approve Depo-Provera until 1992, though the patient label for the lower-dose version approved in

October 2004 comes with a black-box warning that it may lead to a permanent loss in bone density. The warning also specifies a two-year limit on use.[26]

Depo-Provera was, and in many places continues to be, the contraceptive of choice for governments that want to provide an easy-to-use method that does not require maintenance beyond the four-times-yearly injection. Tracing one country's promotion and dissemination of the shot illustrates the ways that governmental support structures the reproductive health options of people dependent on public health-care services. South Africa's use of Depo-Provera in its health-care system from the 1970s forward illustrates how the technological convenience of a long-acting reversible contraceptive (LARC) supported the functions of "everyday apartheid" in citizen's medical care.

South Africa adopted Depo-Provera in 1973 for its public family planning clinics, which numbered 2,045 by 1976. A generation later, in 1992, there were over sixty thousand "clinic points" around the country designed to serve rural, largely black citizens. Every three months, health workers stopped at these clinic points, which provided, among other services, free Depo-Provera shots to women seeking contraception. Other contraceptives and reproductive services were available (including pap smears and STI screening) but only with a cost, and conversations between workers and clients about reproductive health were perfunctory at best. Given that national publicity campaigns regarding the spread of HIV/AIDS did not begin until 1992 and that there was no sex education in the national school curriculum, these clinic visits were a missed opportunity to provide education and reproductive health care beyond just the shot.[27] The British Anti-Apartheid Movement successfully lobbied the IPPF to expel South Africa from its organization, given that Depo-Provera had long-term health implications of which receivers were not fully informed, it caused cancer in animal testing, and its distribution was clearly racialized and directed specifically at reducing the black population.[28]

The end of apartheid in 1994 coincided with the growth of the HIV/AIDS epidemic in sub-Saharan Africa, and the new democratically elected government aimed, among its other goals, to address unmet contraceptive needs and the HIV/AIDS crisis together.[29] Thirteen years after the official end of apartheid, in 2007, the South African government passed legislation that the state should provide all types of contraception to any women or girls twelve years of age or older. According to a 2012 report, injectable progesterone-only contraceptives remain the most common choice in the country, followed by pills and male condoms.[30] The National Department of Health published a revised national Fertility Planning Policy in the same year, which encouraged pairing contraception with HIV/AIDS prevention through the use of a hormonal method alongside a male condom.

However, some apartheid-era problems remain, and new problems following the implementation of antiretroviral therapy (ART) for HIV emerged. According to Diana Cooper and colleagues, "confusion over guidelines for implant implementation, inadequate piloting, limited human resources and training, and pre-emptive concerns about its perceived lower effectiveness in preventing pregnancy in WLHIV (Women Living with HIV) who are on ART have contributed to removals of [Implanon brand] implants and decreased insertions."[31] Ongoing research on the interactions of ARTs (either for therapy or prevention) with hormonal contraception—both medicines that people may take for decades—is necessary to understand how each type of medicine affects the efficacy and safety of the other.[32] Examining both the apartheid- and post–apartheid-era situations regarding contraceptives in South Africa provides a snapshot of the multifaceted challenges facing countries with legacies of colonialism, racism, economic and social inequality, and sexism—traces of which still shape policies, attitudes, and actions in the present.

Emergency Contraception

As research and experimentation with LARCs expanded rapidly in the 1960s and early 1970s, some doctors were thinking about how they could be used on a short-term basis. A. Albert Yuzpe, a physician based at the University of Western Ontario, Canada, had regular requests from students for emergency contraception after unprotected sex. He was reluctant to prescribe diethylstilbestrol (DES) given its severe side effects and sought an alternative.[33] He read a study about a postcoital contraceptive that combined the progestin compound levonorgestrel with ethinyl estradiol (an estrogen) taken within seventy-two hours after unprotected sex, in two doses twelve hours apart. He and an American colleague, Lee H. Schilling, published a study in 1980 reporting the effectiveness of this method using the combination oral contraceptive Ovral, and physicians and health-care practitioners in North America and Western Europe began to prescribe the easy-to-follow "Yuzpe method" off-label. However, the lack of FDA approval, the reluctance of major pharmaceutical companies to produce a specific version of what came to be called "the morning-after pill," and a lack of labeling to that effect made it challenging for patients to trust that the method was legitimate and effective.[34] American pharmaceutical companies that manufactured daily contraceptive pills thought that Americans would confuse the morning-after pill with RU-486 (mifepristone, later combined with misoprostol), the "abortion pill" that became available in France in 1988.

The Yuzpe method garnered attention from feminist health advocates throughout the 1980s and 1990s, particularly those who worked with victims of sexual violence. The morning-after pill could help women who did not want to risk pregnancy after rape. In 1996, seven organizations formed the International Consortium of Emergency Contraception and found a Hungarian industry partner, Gedeon Richter, which agreed to manufacture the pill. However, Richter did not want to distribute or to market the pill in the United States, so the US activist Sharon Camp founded her own company, Women's Capital Corporation (WCC), in January 1997 to put forward an FDA application for "Plan B." It was subsequently approved in 1998.[35] As American companies realized that a market for a morning-after pill existed separately from the market for RU-486, Barr Laboratories purchased the patent rights for Plan B from WCC in 2006.[36] Plan B became available over the counter in the US without a prescription for people seventeen and older in April 2009, though its actual accessibility remains subject to pharmacists' invocation of conscience clauses.[37]

The Male Pill

An ongoing complaint among women using hormonal contraceptives is that there is no pill for men so that they can share in the side effects as well as the pleasure of sex without pregnancy risk. Hormonal contraceptives for men have indeed been an ongoing subject of research for international organizations from the 1950s onward. For example, the WHO's Group on Fertility Control established a Male Task Force in 1968 for research and development of male contraceptive methods but failed to find any workable method without side effects. Laboratories in the WHO's network in India and China also unsuccessfully researched progesterone compounds in the 1980s and 1990s. Pharmaceutical companies were reluctant to get involved in research and development for a male pill because the female pill was highly profitable and they saw no need to split the market for hormonal methods.[38]

However, the interest in a "male pill" has not disappeared. The challenge for scientists who research male contraceptives is that the progestin and estrogen in hormonal contraceptives interrupt spermatogenesis and testosterone production, so testosterone must be added to male hormonal contraceptives to maintain a constant level of it in the body. Alternatively, they can be combined with a substitute androgen to maintain efficacy. Researchers in this area focus on hormonal methods delivered as implants, gels, or syringes because testosterone does not enter the blood in sufficient quantities when taken orally (chapter 6).[39]

Conclusion

A confluence of people and organizations with complimentary skills and interests brought the hormonal contraceptive pill into being in the late 1950s. Further experiments, often off-label, led to new developments in contraceptive use that later expanded the ways that users, health-care practitioners, and population controllers or family planning advocates conceived of the technology. Although national governments and major international health organizations adopted new contraceptives into their planning and activities unevenly into a range of contexts and for various (often racist and classist) reasons, their thinking and motivation for action was usually on a macro level. As this chapter has outlined, however, the contraceptive experience of those targeted for population control was always on an individual level. As Elaine Tyler May puts it, "women eagerly sought birth control wherever it was available. But their motives were *personal*. They used contraceptives to control their own fertility, not to control world population."[40]

Women and people with uteri benefitted individually from advances in hormonal contraceptive technology but also suffered from the often serious side effects that resulted. Moreover, there was no direct line drawn between the pill and the sexual revolution. Nonetheless, the introduction of a contraceptive method that was separate from an act of sex sparked a quiet revolution in many people's reproductive thoughts and actions, whether they embraced a hormonal method, rejected it, or fell somewhere in between. Research on hormonal methods for men continues to draw attention and controversy. Although hormones expanded contraceptive possibilities for many, it did not foreclose the use or further scientific development of others. Chapter 4 revisits the nonhormonal methods outlined in chapter 2 and traces changes in those technologies and the contexts in which people chose them.

NONHORMONAL CONTRACEPTION AFTER THE PILL

The introduction of the pill shifted the context and meaning of the contraceptive methods that were available beforehand. Increasingly after 1960, national and international health organizations centralized and globalized the research, distribution, and testing of contraceptives. Although some users continued with the same methods, many of those who had access to hormonal methods gave them a try. Some users decided to stay with a hormonal method, others returned to previously established nonhormonal methods, and yet others were drawn to historical methods reformulated or repackaged to fit the needs and interests of newer generations. Sometimes users who were unsatisfied with available technologies decided to create and to market their own. No historical contraceptive disappeared completely after the pill, including those forced on people without their consent. This chapter addresses each of the methods first described in chapter 2 and examines how they changed after the pill.

IUDs

The IUD had limited availability before the pill, but the pill's release inspired other scientists and users to revive the technology on a wide scale. As concerns grew over high blood pressure and thrombosis resulting from pill use and as rhetoric about the need for population control became stronger, scientists and users themselves began experimenting with the old form. In Cuba, the US embargo that began in 1962 meant that American-made contraceptives, such as condoms and diaphragms, became difficult to obtain. An anthropologist interviewed a sixty-six-year-old Cuban woman in the 2000s who remembered "contraceptive parties" where women made their own IUDs out of fishing wire at Federation of Cuban Women meetings. In the 1980s, Soviet-made IUDs became available, but they caused many more side effects than the homemade ones did.[1]

On a much wider scale, contraceptive research and development in order to curb population was a central aim of many US and international nonprofit organizations, including the Population Council, the World Health Organization (WHO), the United Nations Population Fund, and the US Agency for International Development (USAID). The Population Council sponsored clinical trials of IUDs in a variety of shapes in Germany, Israel, Japan, and Chile. The physicians Jack Lippes and Lazar Margulies designed plastic IUDs that they tested at the University of Michigan Medical School from July 1964 to July 1965, though Hugh J. Davis designed what would become the most popular IUD, the Dalkon Shield.[2]

Throughout the 1960s, the US Food and Drug Administration, concerned by reports about harm due to the pill, approved many kinds of IUDs for sale as pill alternatives through its Office of Medical Devices. Davis wrote the introduction to Barbara Seaman's revelatory *The Doctor's Case against the Pill* in part to promote the Dalkon Shield as the best of them. By January 1968, approximately 3 million IUDs were in use, and by March 1971, over seventy different brands were on the market.[3]

As a result of the public interest in pill alternatives, the medical device company A. H. Robins bought the Dalkon Shield design from Davis's Dalkon Corporation. After Davis made some modifications to the design in October 1970 that were never clinically tested, A. H. Robins began manufacturing it in January 1971. Soon after doctors began inserting them, however, their problems became clear. Davis had hidden the worrisome results of his limited device testing, so the FDA was not initially aware that the product's multifilament tail string was an ideal site for bacteria to grow and to wick into the vagina and uterus. The fishhook design, which was supposed to reduce the rate of expulsion, instead caused uterine perforations, sepsis, permanent infertility, ectopic pregnancies, and even (in twenty cases) death. Production continued until June 1974, and the devices were still distributed until April 1975. By the time production finally ceased, 2.2 million Dalkon Shields had been sold in the United States, and 2 million more unsterilized Shields had been shipped to seventy-nine countries. The official FDA product recall was not issued until October 1984, and by then, other brands of IUDs had been recalled, as well. Robins had to settle 197,000 damage claims from women who had had the Dalkon Shield inserted.[4]

Parallel to the Dalkon Shield fiasco, the Population Council also supported research and development of a copper IUD with monofilament strings in Chile, designed for women who had not been pregnant. The FDA approved pharmaceutical company Searle's copper-T IUD, also known as the CU-7, in 1976. It also caused serious problems such as pelvic inflammatory disease and infertility and was withdrawn from the market in 1986. A new generation of the copper-T, the copper-T 380, was FDA-approved in 1984 and remains in production under the brand name Paragard. An IUD called Mirena that uses the hormone levonorgestrel instead of copper as the preventative agent has been FDA-approved since 2002.[5] Both

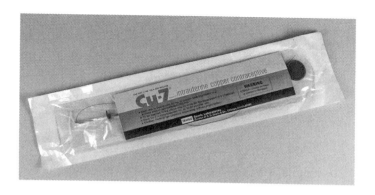

Figure 13 Cu-7 Intrauterine Copper Contraceptive, G. D. Searle, Chicago, IL, I.L., ca. 1976. *Source:* Courtesy Medicine and Science Collections, National Museum of American History, Smithsonian Institution, Washington, DC.

the hormonal and non-hormonal IUDs are marketed to women in committed heterosexual relationships who do not require additional protection against STIs.

The Dalkon Shield disaster and ongoing concerns about the strength of the pill impacted initiatives and policies for American women's reproductive health in two ways. First, the FDA sought to correct its mistakes by strengthening its medical device regulations. After the Medical Device Amendments were passed in 1976, the agency had more power to act when devices were found to be hazardous. The Medical Device Amendments reclassified the approval regulations for all medical devices under three classes according to the level of federal oversight required: class III devices were riskiest and had the most regulations, and class I devices were the mildest and had the fewest regulations. Second, the feminist women's health movement saw an opportunity to reintroduce a historical technology, the cervical cap, to the American market, and they turned their energies to doing so.

Diaphragms, Cervical Caps, and Female Condoms

A barrier method that equaled the male condom for efficacy, safety, low cost, ease of use, and STI protection was of great interest to birth controllers in the first half of the twentieth century and to population controllers and reproductive health advocates in its second half and beyond. The diaphragm, cervical cap, and female condom all waxed and waned on the global market on account of the shifting foci of feminist activism, levels of support from nongovernmental organizations and private foundations, national and international device regulations, and shifts in user needs, interests, and motivations.

In addition to the preface recommending the Dalkon Shield, *The Doctor's Case against the Pill* also included promotion of the diaphragm and cervical cap as pill alternatives.[6] The soon-to-be-classic "Women and Their Bodies" mimeographed pamphlet (later *Our Bodies, Ourselves*) produced by the Boston Women's Health Book Collective in

1970 mentioned these barrier methods as well but pointed out the difficulty of obtaining them. As a result of these two publications, American women began asking their gynecologists about the cervical cap and were stymied by a general lack of knowledge regarding the device. The feminist women's health movement, in which Seaman played a central role, began to establish volunteer-run medical clinics devoted primarily to women's reproductive health across the country and established the Feminist Women's Health Network (FWHN) in 1975 (chapter 3). These clinics initially focused on providing clients with information about contraception, abortion, and reproductive health and opportunities for group discussion and consciousness raising. It was soon clear that women wanted not only information but also the technologies that would help them contracept pill-free as well.

Feminist health clinics in Concord, New Hampshire, and Iowa City, Iowa, began importing the caps individually from the only remaining manufacturer, Lamberts Dalston Ltd. in England, because no domestic manufacturers existed. A challenge for the feminists to expand cervical cap distribution was the 1976 Medical Device Amendments' class III restriction that permitted only physicians to receive and to distribute devices that had not met specific safety requirements. It was not long before US Customs confiscated a shipment of cervical caps to a lay-run women's health clinic in Los Angeles. Feminist health clinics

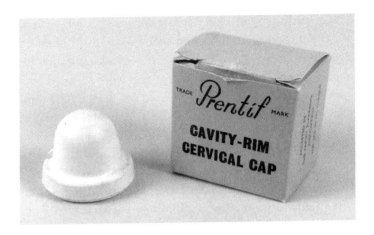

Figure 14 Prentif Cavity Rim Cervical Cap, Lamberts Dalston, Ltd., London, ca. 1979. *Source:* Courtesy Medicine and Science Collections, National Museum of American History, Smithsonian Institution, Washington, DC.

needed a different strategy to get cervical caps to clients, so they decided to participate in a National Institute of Child Health and Human Development (NICHD) call for contraceptive device trials beginning in 1980.[7]

Inspired by the success of the NICHD trial and patients' largely positive experiences, a larger group of women's health centers decided to take the next step in obtaining approval to distribute the three kinds of cervical cap that Lamberts produced, which was to complete an investigational device exemption (IDE). An IDE allowed them to test the device on a larger sample of volunteers.

Although many patients appreciated that the cap was hormone-free, was reusable for a year or two, and could be inserted hours before sex, it leaked and smelled after removal, their partners' penises banged into it and became bruised, and it occasionally dislodged during vigorous or prolonged sex. The FDA rejected one brand of cap, the Vimule, in 1984 due to small lacerations on the cervixes of women volunteers, but another, the Prentif, was approved in May 1988. For unclear reasons, however, Lamberts chose a single US distributor unaffiliated with the FWHN, and after some rancorous back-and-forth correspondence with her, the network abandoned the decade-long project in 1991. A silicone cervical cap called FemCap, unrelated to Lamberts, has been available in Europe since July 1999 and in the US since March 2003.[8] Although the FWHN was unsuccessful in bringing cervical caps to women at its health clinics, it was successful in bringing public attention to the technology so that others could later capitalize on the market for woman-controlled barrier methods.

The diaphragm, which was larger than the cervical cap, also reemerged. The Atlanta, Georgia, Feminist Women's Health Center was involved in testing a silicone version for nulliparous (never-before-pregnant) women called Lea's Shield in the 1990s. The device had a hole in the middle, called a flutter valve, supposedly in order to allow drainage of cervical secretions, though it created an obvious design flaw that limited its utility. The US FDA approved it in 2002, but the manufacturer discontinued it in 2008. A silicone diaphragm called Caya, manufactured in Germany, is now obtainable in many countries either with or without a prescription. Like the cervical cap, using it with spermicide improves its efficacy.[9]

Yet another barrier method, the female condom, became a focus of the international public health community in the late 1980s. A rare mention of a female condom before then was in Marie Stopes' 1924 book *Contraception (Birth Control): Its Theory, History, and Practice. A Manual for the Medical and Legal Professions*, where she noted the device's benefits: "they do avoid the personal discomfort ... which so often leads to a reduction of [a man's] erection after applying the condom himself. Sometimes a woman is aware of her husband's contamination with venereal disease. ... such an unfortunate wife should certainly use this protective sheath."[10] Although female condoms may have improved men's erections and decreased disease transmission, they could cover the clitoris depending on length and positioning, lessening women's erotic sensations. It took time during foreplay to put them on, they required spermicide, they were expensive compared to male condoms, and they were challenging to obtain outside of England.

Similar pro and con arguments appeared when the female condom reemerged at the end of the 1980s, largely as the result of NGOs seeking a woman-controlled contraceptive that could be manufactured and distributed in the developing world easily and inexpensively. A short-lived product called the Bikini Condom, disposable latex underpants with a flexible vaginal pouch, appeared and then promptly disappeared in 1991. A longer-lived polyurethane version was released by a Danish physician, Lasse Hessel, in 1990, and the company Wisconsin Pharmacal (later the Female Health Company) followed with its version (called "Reality") a few years later. Repeated complaints were that they smelled bad and squeaked during intercourse. The second-generation version, FC2, now manufactured by Veru Inc. and consisting of a nitrile sheath and outer ring, received FDA approval in 2009. In the United States, they received a reclassification from a class III to a class II device in October 2018, and their official designation changed from "single-use female condom" to "single-use internal condom." This shift in classification indicates that this technology no longer needs premarket notification requirements, has proven safety and effectiveness, and can be sold over the counter. Moreover, the name change makes the technology more trans-inclusive, centering on its placement instead of the user's gender.[11]

Figure 15 FC2 Female Condom, ca. 2018. *Source:* Reproduced with the kind permission of the Female Health Company, Veru Inc., London, England.

Condoms

Basic condom technology has undergone few changes over the years, though the availability of different sizes, qualities, lubrications, spermicides, colors, textures, and thicknesses has expanded significantly. Condoms gained further lasting significance in the 1980s not just in preventing pregnancy but also as "life-saving protection against disease transmission" in homosexual communities at a peak of the HIV/AIDS crisis.[12] Although the object has remained the same, the national contexts in which it is imported, manufactured, advertised, sold, or given away and the reasons for its use differ according to national laws, health-care regulations, levels of religiosity, and contraceptive cultures. To illustrate those differences, this section examines the history of condoms from the mid-twentieth century to the present in three countries: Japan, Ireland, and Uganda.

During and after the Allied occupation of Japan following World War II, the country's leadership aimed to stabilize the country in part through promoting contraceptives, lowering abortion rates, and combatting poverty. The 1948 Pharmaceutical Law explicitly allowed the sale of condoms and diaphragms, and the prime minister's cabinet council established a national policy in October 1951 to popularize contraception and to educate birth control instructors. The leader of the National Institute of Public

Basic condom technology has undergone few changes over the years, though the availability of different sizes, qualities, lubrications, spermicides, colors, textures, and thicknesses has expanded significantly.

Health, Yoshio Koya, had support for his research from the Population Council and contraceptive advocate Clarence Gamble. Gamble pushed foam powder with sponges and spermicidal jellies, but Koya found that most people he interviewed preferred condoms: their use among married women ages sixteen to forty-nine rose from 35.6 percent in 1950 to 68.1 percent in 1969.[13]

That preference remained stable because of a lack of other options and national protection of the condom industry. Four major companies and around a dozen smaller ones produced 35.5 percent of the world's total condoms in 1968, and Japanese brands cultivated a worldwide reputation for very thin (as little as 0.01 millimeters) and durable polyurethane condoms that heightened sensation during intercourse. One result of the condom industry's dominance was that the Central Pharmaceutical Advisory Council of the Japanese Ministry of Health and Welfare approved the hormonal pill for sale only in June 1999, thirty-nine years after its release in the United States. The copper-T IUD also received federal approval the same year, fifteen years after its US debut. A 2010 article noted that many Japanese women disliked the pill because of bloating, weight gain, and the need for a prescription; those who used it usually took it for other purposes, such as menstrual regulation. A 2016 survey of married Japanese women ages sixteen through forty-nine indicated that their rate of condom use was 83.4 percent and pill use was

3 percent.[14] Japanese condoms can be purchased presently in a dazzling array of sizes, colors, flavors, and thicknesses anywhere from "skin corner" kiosks to multistory department stores. The decades-long illegality of the pill and IUD supported innovation and creativity in the condom industry while limiting the visibility and desirability of contraceptive alternatives.

In the Republic of Ireland, one was hard-pressed to find condoms or any other contraceptives for nearly sixty years. The 1935 Criminal Law Amendment Act criminalized the sale and importation of contraceptives in order to align with the Catholic Church's prohibition against artificially limiting births. In 1973, however, contraceptives were constitutionally allowed for married couples, and the 1979 Health and Family Planning Act confirmed that married people could obtain contraception from pharmacists with a doctor's prescription. In 1985, the act was amended to allow sales to anyone without a prescription, though sales were limited to pharmacies, medical clinics, and doctor's offices. Contraceptives were legalized fully in 1992, in part as a delayed response to the country's AIDS crisis.[15]

Decades of activism paved the way to contraceptive legality. Members of the Irish Women's Liberation Movement organized a one-time "contraceptive train" on May 22, 1971, from Dublin to Belfast, Northern Ireland, where contraceptives were legal. They had decided

In the Republic of Ireland, one was hard-pressed to find condoms or any other contraceptives for nearly sixty years.

to draw attention to their "Chains or Change" document that made six demands of the Irish government, including contraception, equal pay, and equality before the law. They bought condoms, diaphragms, IUDs, hormonal pills, and spermicidal jelly and brought them back to Dublin with the intention of provoking customs officers at Connolly train station to arrest them. On governmental orders, the officers refused and let the women through. The action did not change the law, but the women's collective efforts highlighted the absurdity of the national prohibition against contraception.[16]

In addition to the one-time contraceptive train event, Family Planning Services Ltd. (later the Irish Family Planning Association) established Well Woman clinics in the 1970s, and radical feminists backed a Contraceptive Action Programme (CAP) that set up temporary contraceptive shops in Dublin and Cork. The Well Woman clinics skirted the law prohibiting import of condoms for sale by giving them away for free but soliciting donations in support of their educational activities. Both groups included Protestants who were accustomed to circumventing the Catholic Church on other matters. The final push for legalization came in the early 1990s from two sources—a group called Condom Sense, which persisted in setting up condom vending machines in the nonmedicalized spaces of bars and clubs in 1992, and an Irish Family Planning Association booth in the Dublin Virgin Megastore, which deliberately provoked legal action by selling Mates condoms to a police detective in January 1990. The illegal activity of these groups "established new Irish modes of engagement with contraception, not yet provided by the state, which were no longer saturated by religious morality, or, necessarily, by conservative medical power, but instead were characterized by solidarity with clients, care, and even humor."[17] Between 1971 and 1992, the condom had shifting symbolic value in Irish public culture—first as a tool of family planning, then as a central element of the fight for reproductive rights, and finally as an emblem of sexual wellness and the ability to enjoy sexual pleasure with less fear of STIs.

Condoms play a different role in Uganda, influenced by a long tradition of missionary presence, deep-rooted homophobia, and conflict over gender roles in public and private settings.[18] Family planning organizations began distributing condoms in the 1950s, but that distribution was never comprehensive. As a result of many Ugandans' conservative Christian beliefs, long-standing antihomosexual beliefs (including an invalidated fall 2009 bill that would have made some forms of homosexuality punishable by death), and the promotion of abstinence-only programs by President Yoweri and First Lady Janet Museveni, condoms are difficult to obtain regularly throughout most of the country. The country also has high fertility and HIV rates: 6.77 children per woman in 2017, with 4.5

percent of men and 7.3 percent of women testing positive for HIV.[19] In theory, many people would be interested in limiting exposure to disease and spacing pregnancies, but condom uptake is low.

Focused interviews with men and women about the condom in southwest Uganda in the 1990s illustrate how a single object encapsulates multiple meanings and why NGO and public health attempts to extend and to stabilize the market for condoms are often unsuccessful. Many women did not trust condoms or their partners to use them correctly. Some thought that condoms could fall off the penis and become lodged in the uterus and that men put holes in a condom when they wanted to get their partner pregnant without her consent. Furthermore, a woman insisting on a female or male condom would upend traditional gender roles in sexual encounters. Men interviewed believed that condoms were porous and let HIV pass through by either manufacturer manipulation or improper storage. Using condoms in a long-term relationship was tantamount to confessing infidelity.[20] The condom is thus a reminder of the need for protection, a barrier between sexual partners to prevent STIs and pregnancy, a means of controlling one's health, a device signifying a lower fee for sex workers, a symbol of trust showing that one's partner cares about one's safety, and at the same time, a symbol of mistrust and infidelity that either partner could sabotage.

Activists, NGOs, the national Ministry of Health, family planning clinics, and sexually active individuals themselves are aware of all the complexities surrounding the condom. From 2013 to 2015, the Ministry of Health implemented a Comprehensive Condom Programming Strategy that resulted in a final report outlining the status of condom use, distribution, and access across the country. It showed that the national medical stores that handled all essential medicine distribution, the Uganda Health Marketing Group, and private NGOs were all involved in condom distribution. Multiple factors influenced uptake: a lack of education regarding proper condom use; the importation of poor-quality or expired condoms and a subsequent lack of trust in them; inadequate storage space at warehouses and distribution centers; inadequate distribution patterns leading to stockouts in some areas; a lack of access to condoms on Saturday nights, the time of highest demand; a stigma against nonprostitute women carrying condoms; and anticontraceptive propaganda from political and religious leaders.

Furthermore, even though the female condom was introduced in the 1990s, use remains very low due to lack of access to and information about it. The report was clear: "condom use and condom distribution are correlated. The larger the quantities of condoms distributed, the higher the likelihood that more people will use condoms."[21] The report illustrates the complexities of ensuring a consistent countrywide supply of high-quality condoms to users who can access them when they are needed most. The condom in Uganda exists in complicated sets of tug of war between NGOs and for-profit manufacturers, pro- and anticontraception government agencies, progressive and patriarchal gender ideals, and the desire for safety and the need to avoid shame and embarrassment.

Spermicides and Sponges

After nonoxynol-9 was patented in 1951, most researchers interested in spermicides turned their attention to the medium in which the spermicide was suspended. Because spermicides function in both mechanical and chemical ways, finding a suspension and delivery medium that covered the vagina and stayed put during intercourse was critical to a spermicide's effectiveness. The need to conduct spermicide testing in vitro (in a laboratory) instead of in vivo (with human subjects) due to obvious ethical problems if the substance failed hampered scientists' abilities to test substances for efficacy under real human conditions. A few US researchers in the late 1950s and early 1960s tried home testing—for example, asking test subjects to come to a laboratory for testing within four and a half hours after coitus using Emko aerosol foam. Another researcher asked female patients to bring in fresh (one- to two-hour old) samples of sperm into a laboratory; the patient applied the spermicide and masturbated with a penis-shaped tube for two minutes before the doctor "ejaculated" a syringe of her husband's sperm into her; and then the doctor took swabs from different areas of her vagina to see how well the spermicide worked. This researcher noted that "the amount of cervical coverage was at least partially dependent upon the depth, force, and duration of thrust; shallow motions were less apt to disturb the spermicide than deep ones." Furthermore, if sex lasted more than two minutes, changes in vaginal pH and temperature affected the spermicide's staying power.[22]

This latter experiment highlighted an obvious yet understudied problem in contraceptive research: how people have sex impacts the effectiveness of barrier contraceptives. That problem was highlighted in research conducted by William H. Masters and Virginia E. Johnson before they became widely known through the publication of *Human Sexual Response* in 1966. At the behest of the National Committee on Maternal Health, Masters and Johnson tested spermicides infused in foams, creams, and gels on women who either masturbated with a dildo camera to orgasm at various angles or had sex in multiple positions with their real-life partners. They found that Delfen Cream and Emko Foam provided the best coverage—and that coverage worked best in the missionary position with enough foreplay before intromission to ensure that products distributed evenly. The testing also revealed that having sex in multiple positions, lengths of foreplay and intercourse, women's level of natural lubrication, the number of children the woman had previously, and the use of artificial lubricants all affected contraceptive effectiveness.[23]

Other scientists in the 1970s experimented with alternate delivery mechanisms that had little staying power. English and US researchers tried C-Film, a film that dissolved in the vagina thirty minutes before intercourse,

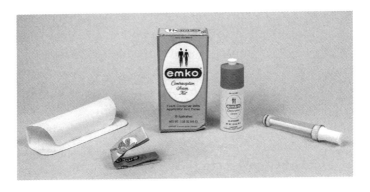

Figure 16 Emko Contraceptive Foam Kit, Schering Corp., Kenilworth, NJ, ca. 1965. *Source:* Courtesy Medicine and Science Collections, National Museum of American History, Smithsonian Institution, Washington, DC.

and WHO scientists investigated a spermicide ring placed inside the vagina in 1979. A company called Nova Corporation tried to market a flavored spermicide called Flavor-Cept designed for use with oral sex. Also in the 1970s and 1980s, laboratory tests showed that spermicides could potentially protect against STIs, including gonorrhea, chlamydia, and HIV. However, WHO scientists in the Department of Reproductive Health and Research found later that nonoxynol-9 in fact irritated the vagina and made HIV transmission more likely.[24] In the end, spermicides remain a moderately effective contraceptive option on their own but are most effectively used in combination with barrier methods.

Such barrier methods included the Today sponge (a disposable polyurethane sponge permeated with spermicide), approved by the US Food and Drug Administration (FDA) in April 1983 and discontinued in January 1995. The manufacturer of the Today sponge, VLI Corporation in Costa Mesa, California, received notification from the FDA in March 1994 that upgrades to its only production facility were necessary. The company could not afford the upgrades and closed the facility instead. VLI's decision became part of 1990s American pop culture, when the character Elaine (Julia Louis-Dreyfus) on the television show *Seinfeld* hoards the soon-to-be-discontinued product and has sex only if a prospective partner is "sponge-worthy." Though intended to be humorous, the episode highlighted

Figure 17   Today Vaginal Contraceptive Sponge, VLI Corp., Costa Mesa, CA, ca. 1983. *Source:* Courtesy Medicine and Science Collections, National Museum of American History, Smithsonian Institution, Washington, DC.

more serious issues of product availability. Several other companies have since attempted to manufacture and to market a similar product. The American brand Mayer Labs currently offers a sponge available at drugstores without a prescription.[25]

Finally, by the 1960s, marketing of suppositories and douches as contraceptives had all but disappeared. American douche manufacturers such as Beecham-Massengill Pharmaceuticals and Morton-Norwich Products, which produce Massengill and Jeneen douches, respectively, marketed their products as providing "feminine hygiene." Present-day douche formulas are less astringent than those in the past but remain just as unnecessary for vaginal health.

Herbs and Heat

The use of herbs for contraceptive purposes decreased but did not disappear after the pill, though professional research and development shifted from women to men. Research on an initially promising herbal contraceptive for men was conducted in China in the 1970s with the support of the Task Force on Methods for the Regulation of Male Fertility of the WHO. Throughout the decade, Chinese investigators tested gossypol, a component of cottonseed oil, on fourteen thousand men across the country. The men took a daily dose of the oil for several weeks, then once or twice a week afterward, and its efficacy reached 99 percent. Unfortunately, its serious side effects (fatigue, potassium deficiency, nausea, dizziness, diarrhea, circulatory problems, heart failure, and permanent sterility) were unsolvable problems and prevented the WHO from sponsoring further clinical studies.[26] Contraception using herbal methods is now rare, and when it occurs, is usually conducted under the guise of menstrual regulation.

The modification of fertile bodies through temperature was another avenue that physicians explored. Because testes must remain several degrees cooler than normal body temperature in order to maintain normal spermatogenesis, some experimented with procedures and devices that would heat the testes and kill sperm as a result. Marthe Voegli, a Swiss physician working in India from 1930 through 1950, advocated a "wet heat" method in which men would submerge their testicles in a shallow bath of 116 degrees Fahrenheit for forty-five minutes a day for three weeks. Even though she declared that this temporary sterility would last six months, when normal fertility would return, few if any men adopted this method. John Rock, best known for his work on testing the hormonal pill in the 1950s, also explored this method in the 1960s. He collaborated with a physician colleague at Harvard University to develop insulated underwear—a jock strap with the hard plastic removed and replaced with oil cloth and tissue paper—that would keep testes at a heightened temperature. They called the method "artificial cryptorchidism," as it mimicked the sperm-killing medical condition in which one or both testes did not descend fully from the body. A French doctor revived the method in the late 1980s through the mid-1990s with his own underwear design, but it, too, generated little lasting public or medical interest.[27]

Timing Methods

More popular than herbs or heat was a renewed interest in timing methods. After the pill's quick adoption worldwide, the Roman Catholic Church hierarchy decided to revisit its position on contraception by establishing a Pontifical

Commission on Population, Family, and Birth in October 1963. Its majority report, which leaked to the public in spring 1967, declared that "the regulation of contraception appears necessary for many couples who wish to achieve a responsible, open, and reasonable parenthood." However, Pope Paul VI decided to ignore the Pontifical Commission's report in the encyclical *Humanae vitae* in July 1968. The encyclical rejected the Pontifical Commission's recommendation to allow married couples to use the pill and instead reaffirmed its position that "any action which either before, at the moment of, or after sexual intercourse, is specifically intended to prevent procreation" was forbidden. It also rejected voluntary and involuntary sterilization. *Humanae vitae* included a positive suggestion for faithful married couples: it encouraged sex during a woman's infertile period. At first, the church advocated the "rhythm method" but later focused on natural family planning (NFP).[28]

As news of the encyclical ricocheted around the world of Catholic believers, clergy members, doctors, and the faithful, they had to decide whether they would adhere to and promote its unexpected directives. Many had expected that the encyclical would approve pill usage and proceeded in the seven-year period between the pill's wide availability and the encyclical's promulgation to encourage pill use among the faithful. For example, Juan Landázuri, the cardinal of Lima, Peru, favored the pill at first and supported its testing at eight experimental clinics linked to local parishes, but his support gradually waned after *Humanae vitae*. Some Catholic physicians supported the pill to help regulate the menstrual cycles of women who wanted to use timing methods with more accuracy. Nevertheless, by the late 1970s, any remaining support for the pill from Peruvian Catholic leaders had withered away. Latin American cardinals ended up supporting the pope's decision, at least publicly, but striking students at the Universidad del Rosario in Bogotá, Colombia, staged protests satirizing the encyclical in front of the medical faculty and were thrown out of the university. A longitudinal survey of Catholic women worldwide from 1960 to 1987 illustrated that *Humanae vitae* had little to no effect on whether they used the pill.[29] There was additional unevenness in the worldwide church regarding its replacement methods. The rhythm method, which predicted fertility algorithmically based on a woman's past menstrual cycle and basal body temperature, was relatively accurate post-ovulation every month but less so preovulation.

To improve the accuracy of timing methods, the Catholic Church now promotes the three methods of NFP that should be used together: observation and charting of cervical mucus, sympto-thermal methods (charting basal body temperature to pinpoint the timing of ovulation), and sympto-hormonal methods (self-detection of reproductive hormones in the urine). A half degree rise in a woman's body temperature indicates that ovulation has occurred. Her cervical fluid changes in texture, and her cervix alters in texture and shape. According to the US Conference of Catholic Bishops, "NFP is unique among methods of family planning because it enables its users to work with the body rather than against it. Fertility is viewed as a gift and a reality to live, not a problem to be solved. The methods of Natural Family Planning respect God's design for married love!"[30]

Whether part of a divine plan or not, timing methods are the province not only of faithful Catholics alone but also of those interested in managing their fertility without hormones or barrier methods. The fertility awareness method (FAM) uses techniques similar to NFP (mucus, temperature, and hormonal tracking), but its adherents practice it for markedly different reasons. Its lack of forbidden barriers or chemicals attracts the religious, but FAM's lack of chemicals attracts those interested in keeping their bodies free of artificial hormones. NFP and FAM are limited to a small group of users who are willing to limit intercourse according to monthly "safe" periods; who trust their partners not to cheat or to break the timing rules; who are not at risk of STI transmission; and who have no medical conditions that interfere with the regularity of menstruation, such as hypothyroidism or polycystic ovary syndrome. Those practicing FAM instead of NFP still abstain from hormonal birth control but also use barrier methods if they want to have sex on fertile days. The best-known English-language promoter of FAM, Toni Weschler, argues that the method improves its users' sex lives: "If her partner participates in her charting, she will probably be more sexually responsive. In essence, through his actions he can show her how respectful he is of her body and comfort, and how much he wants to share in the responsibility of contraception."[31]

Despite such a pronouncement, the technology available to manage those methods has undergone changes that make tracking more private. In the past five years, as using smartphones to handle daily tasks has become ubiquitous, companies have developed apps to help people track fertility with a handheld device and a thermometer. They include Conceivable, Glow, Natural Cycles, and Daysy, the latter two of which track a user's daily temperature reading via a branded thermometer. OvaCue, which Wechsler promotes, includes a monitor for saliva and cervical mucus. Despite the pastel-colored websites with assurances of quality control and high rates of effectivity and the US FDA approval of Sweden-based Natural Cycles for women over eighteen in August 2018, users have had unplanned pregnancies even when following the app's instructions precisely. A British journalist expressed her failure with the app by describing the perfect user: "I now know that the ideal Cycler is a narrow, rather old-fashioned category of person. She's in a stable relationship with a stable

As using smartphones to handle daily tasks has become ubiquitous, companies have developed apps to help people track fertility with a handheld device and a thermometer.

lifestyle. ... She's about 29, and rarely experiences fevers or hangovers. She is savvy about fertility and committed to the effort required to track hers. I could add that her phone is never lost or broken and she's never late to work. She wakes up at the same time every day, with a charged phone and a thermometer within reach."[32] Regardless of whether a user's motivations are religious, health-based, or otherwise, using natural family planning methods requires a level of vigilance, personal health, lifestyle regularity, and minimal travel that narrows the chances of the method's success.

Sterilization

Although sterilization is now available legally in many countries as a voluntary, permanent contraceptive method and has a near 100 percent effectiveness rate, people were and are often coerced into using it. The first part of this section highlights how sterilization programs were enacted in three countries—the United States, India, and Peru—to serve political agendas. Sterilization is a harsh tool of governments that wanted to limit childbearing by people of color, people of lower castes and classes, and poor people and, in China, to enforce the one-child, one-family policy, particularly in rural areas.[33] For those who chose the option voluntarily, especially those with uteri, there were problems with newly developed procedures as well, and the second part of this section reviews them.

First, in the United States, the advocacy group Women of All Red Nations estimated that around 42 percent of Native American women were sterilized between 1968 and 1982, with a rate up to 80 percent on some reservations. The Native American Women's Health Education Resource Center in Andes, South Dakota, pointed out that "many native women turn to tubal ligation due to a lack of other options or to the limited nature of those options within the Indian Health Service." In the 1970s and early 1980s, African American, Puerto Rican, and Mexican American women were also targets of state sterilization programs while being refused abortion funds (chapter 5). Medicaid, the federal health-care program for low-income Americans, covered the costs of sterilization but not abortion. Although the safety of laparoscopic techniques has improved over time, too often non-native-English-speaking women or women with low literacy were coerced into consenting to sterilization under confusing guidelines that they did not fully understand.[34]

Second, coerced sterilization also became a signature element of the Emergency in India, which lasted from June 1975 to March 1977. Indira Gandhi and the Congress party proclaimed a national state of emergency in order to manage threats to her and her political party following a guilty verdict in an electoral malpractice case. Instead of resigning, she established authoritarian rule and suspended civil rights so that "coercion and compulsion were explicitly authorized by the central government." Population control was a key part of the civil rights crackdown, though sterilization programs that had been in effect since 1966 had already affected 18 million people. Karan Singh, the minister of health and family planning, established a national population policy in April with sterilization targets for each state in April 1976. Some state governments authorized cash payments, often a month's wages, to men agreeing to sterilization according to the number of children that they already had, and others required sterilization for anyone using public services such as driver's licenses, firearm permits, and bank loans. In 1976, there were 8.25 million sterilizations (6.2 million vasectomies, 2.05 million tubectomies) and an additional 5.75 million in 1977. These operations were far from safe: 1,800 families claimed restitution after the deaths of loved ones due to surgery.[35]

Both men and women were targeted for sterilization before and during the Indian Emergency, but women were the primary target for the sterilization programs administered under former Peruvian president Alberto Fujimori's dictatorship. Fujimori's goal was to reduce the birthrate in order to reduce poverty. As in the US and India, those forced to submit to sterilization were not informed of alternative contraceptive options or the potentially deadly risks of the surgery, and they were often forced to consent during pregnancy or immediately after giving birth. Thus approximately 261,000 mostly indigenous Quechua-speaking women were involuntarily sterilized between 1996 and 2000. Two thousand of the forced sterilization cases were reported to the Peruvian government, and the Inter-American Court of Human Rights confirmed them to be violations of human rights. Fujimori is currently in prison for corruption and human rights violations, but his punishment does not restore victims' fertility.[36]

Those voluntarily seeking sterilization, often as an alternative to the hormonal pill and IUDs, encountered another set of problems. Most US hospitals, if they did not forbid the procedure entirely based on their Catholic affiliation, had an age-parity rule, or a 120 rule (sometimes a 150 or 175 rule). The age-parity rules stipulated that no woman could be sterilized unless the number of her children multiplied by her age was 120 or higher. In the late 1960s and early 1970s, a coalition of population-control organizations (Association for Voluntary Sterilization and Zero Population Growth) and state-level American Civil Liberties Union (ACLU) affiliates conducted a series of lawsuits challenging sterilization laws. Most public hospitals withdrew their age-parity laws following the *Hathaway v. Worcester City Hospital* appellate court decision in 1973, extending the reproductive freedoms that *Roe v.*

Wade and *Doe v. Bolton* established two months earlier. Restrictions in religious hospitals were unaffected, however, and restrictions requiring a husband's consent took longer to fall by the wayside in public medical facilities.[37]

Women in Colombia had a different perspective on the procedure than US women did. Because men usually failed to take responsibility for contraception, women's desire for a method that they could control turned sterilization into the most popular method in the country from the 1970s through the 1990s. They were able to access sterilization from doctors who had traveled to Johns Hopkins University in Baltimore, Maryland, in the early 1970s to learn laparoscopic techniques. Though sterilization was officially forbidden to Roman Catholics, Colombian doctors were inclined to perform the procedure anyway in the interest of their patients' health and desires to control their fertility.[38]

Further, although a vasectomy is a simple surgical procedure that can be performed on an outpatient basis without general anesthesia, sterilization options for women, except for some laparoscopic procedures (fallopian tube and/or ovary removal), continue to require general anesthesia and open abdominal surgery. Beginning in 1981 with the titanium and rubber Filshie clip in Nottingham, England, inventors and medical device and pharmaceutical companies have designed and brought to market a range of implanted devices. The Filshie clip has been used in tens of thousands of sterilizations in Australia, Canada, Ireland, and the United Kingdom, but in the United States, medical literature and consumer reports since 1996 indicate that it has caused tissue necrosis. They and other implanted devices such as the Hulka-Clemens clip, the Falope ring, and the Silastic ring have dislodged, migrated to and embedded in other parts of the body, and expelled vaginally—sometimes decades after implantation.[39]

Despite ongoing problems and continual reports of involuntary device implantation during sterilization operations, implanted-device methods continue to be developed. In 2002, the US FDA approved a method called Essure that the pharmaceutical company Bayer designed for women between twenty-one and forty-five years old. Essure was a nickel-titanium alloy coil placed into the fallopian tube through the uterus, causing scarring that blocks the tubes. The scarring process took three months and needed a special X-ray to determine its completion. Approximately 750,000 women had undergone the surgery by 2015, though its side effects led Bayer to discontinue the procedure in the United States as of December 31, 2018. The company stated on the device's homepage that "the Essure business is no longer sustainable," which was likely in part because Canada, Finland, the Netherlands, and the United Kingdom had already forbidden sales. Those side effects, which could manifest as a result of previously unknown metal or nickel allergies, included mild to moderate pain, cramping, vaginal bleeding, headaches, nausea, vomiting, dizziness, fainting, and ectopic pregnancy.[40]

Sterilization remains problematic for people with uteri. The history of involuntary sterilization affects sociocultural and medical perceptions of voluntary requests for the operation. Even if the operation is chosen freely, many of the technologies and methods that have been used since the 1980s have long-term health risks. As a group, the present-day short- and long-term risks are called postimplant or postimplantation syndrome—problems that those interested in pursuing the method must take seriously.

Conclusion

After the pill appeared on the world market and its side effects were publicized, it catalyzed research in and advocacy of other forms of contraception, including the IUD, the cervical cap, the diaphragm, timing methods, and voluntary sterilization. The pill challenged nations' visions of what contraceptive methods, if any, were best for their citizens and spurred them to act. Although some national-level decisions, regulations, and laws regarding contraception harmed citizens (involuntary sterilization and the Dalkon Shield most prominently), decisions at the local public and individual levels (such as ongoing citizen activism in

Ireland) indicated a grassroots passion for helping others with their contraceptive needs.

At the same time, the pill's introduction played a more minor role in places where condoms' STI protection was just as important as their ability to control fertility, if not more so. Many policies and directives about contraception took place at high levels of international and national governance, but the effectiveness of any contraception came down to the individual user's facility, access, and determination—plus a partner's agreement. There was still no one universally agreed-upon method that was effective, cost-efficient, safe, and easy to use for all, and while that remained the case, activists, theorists, and historians turned their attention to articulating a framework to evaluate actions moving forward—reproductive justice.

CONTRACEPTION IN THE REPRODUCTIVE JUSTICE FRAMEWORK

The development of the concept of reproductive justice provides a significant set of intellectual and practical tools for reframing the past, present, and future of contraception. The scholar-activists Loretta Ross and Rickie Solinger state the premise of reproductive justice clearly: "The right to reproduce and the right not to—the right to bodily self-determination—is a basic human right, perhaps the most foundational human right."[1] Reproductive justice has its roots in post–World War II international human rights activity and the globalization of development aid, along with the civil and women's rights movements of the 1960s. Reproductive justice as a specific iteration of human rights emerged in the 1990s through cooperation between reproductive rights groups organized by feminists of color. Using texts from movement organizers and historians, this chapter traces the origins and current manifestations of the American version of reproductive justice and outlines its importance in the history of contraception. The reproductive justice framework is critical to present-day thinking about contraception because it links the provision and availability of freely chosen contraception to other issues related to health and safety. The framework also provides both inspiration and concrete guidance to individuals, NGOs, international and national nonprofit agencies, and for-profit medical device and pharmaceutical companies that produce and sell reproductive technology.

The Meaning of Reproductive Justice

Reproductive justice is not just a set of ideas to study from an academic distance. Rather, it is designed to connect diverse reproductive health issues in an easily understandable way; to provide guidelines for developing and executing policy for public agencies and NGOs at local, national, and international levels; and to give activists guidelines and talking points for advocating the importance of their work to the public and to public officials. In short, reproductive justice is "a theory, a practice, and a strategy that can provide a common language and broader unity in movements for women's health and rights."[2] Reproductive justice has three primary organizing principles: "(1)

the right not to have a child; (2) the right to have a child; and (3) the right to parent children in safe and healthy environments." Furthermore, "reproductive justice clarifies the need for protection from coerced sex and reproduction and also from coerced suppression or termination of fertility."[3] Thus, a society oriented around reproductive justice supports interconnected policies and principles, including sex education appropriately designed for children, teenagers, and adults; the freedom to choose when or if to have children; access to information about contraception and the technology itself; freedom from child marriage and forced pregnancy; quality prenatal care and birth support; nutritious and affordable food; and the enforcement of health and safety measures for all children brought into the world. These basic principles of reproductive justice are simple enough to state but challenging (not to mention time-consuming and expensive) for governments at any level to enact, even if there was strong political will to do so. However, linking these elements under the broad banner of "reproductive" provides both specific foci for individuals and groups to address and an ideal, inspirational vision of a healthy human society to work toward.

Reproductive justice does not operate independently of or aside from the organizational principles that structured the movements related to making sex and reproduction healthier and safer earlier in the twentieth century.

It both criticizes and builds on concepts of reproductive health and rights that were articulated in the early and mid-twentieth century. These concepts, as outlined in previous chapters, identify the importance of protecting the health of people of reproductive age and providing them access to information and technologies that help them manage their sexual and reproductive lives healthily and safely. Nonetheless, as is abundantly clear, some motivations—population control, the limiting of the reproduction of people deemed "unfit," and the development of new contraceptives—were often stronger for those involved in developing contraceptive technologies historically than ensuring safety, sexual health and fulfillment, or the absence of pain and discomfort. As Loretta Ross writes, "Reproductive justice is a real and present embodied activism by women of pushing against a conservative, racist, and misogynist antisex society that devalues our lives, our partners, and our children."[4] Reproductive justice builds on the most positive legacies of reproductive health and rights, demonstrates the inadequacies and failures of these concepts, and offers a holistic perspective on how to move thought, advocacy, and practice forward in the twenty-first century.

The working group Asian Communities for Reproductive Justice (now Forward Together) clarified the differences and similarities between these three concepts in a 2005 document:

These three conceptual structures together provide a complementary and comprehensive response to reproductive oppression as well as a proactive vision. ...

> *Reproductive Health* is a framework that looks at service delivery and addresses the reproductive health needs of individual women. ...
>
> *Reproductive Rights* is a legal and advocacy-based model that is concerned with protecting individual women's legal rights to reproductive health care services, particularly abortion. ...
>
> *Reproductive Justice* is a movement-building and organizing framework that identifies how reproductive oppression is the result of the intersection of multiple oppressions and is inherently connected to the struggle for social justice and human rights.[5]

Thus, reproductive justice is an intellectual and practical means for identifying the ways that historical and political forces have deprived those capable of reproduction of their abilities to control their reproductive futures. More specifically, it does not just point out past and current inadequacies in reproductive care but points to connections with other human rights movements and provides a vision for enacting just and inclusive reproductive health for all.

The Historical Roots of Reproductive Justice

The history of reproductive justice has three strands: global human rights standards, women's and civil rights movements, and intersectional theories used for analyzing human behavior and sociopolitical systems. Together, these three strands of thought and action have shaped its present-day form.

First, "many feminists around the world prefer to use international human rights standards to make claims for full reproductive freedom," instead of an individual rights–based or privacy-based set of standards.[6] In that way, they can unite reproductive justice to the most fundamental global standards for human rights: the Universal Declaration of Human Rights (UDHR), proclaimed by the United Nations General Assembly on December 10, 1948. The UDHR was drafted in the aftermath of World War II and serves as a centralizing, unifying document around which all nations could unite and declare that the atrocities of that war would never happen again. The articles of particular importance to reproductive justice include article 1 ("All human beings are born free and equal in dignity and rights"), article 16(1) ("Men and women of full age, without any limitation due to race, nationality, or religion, have the right to marry and to found a family"), and article 25(2) ("Motherhood and childhood are entitled to special care and assistance").[7]

By basing reproductive justice on a globally accepted text—the foundation of human rights law worldwide—reproductive justice advocates expand on the rights enumerated in the UDHR specifically for reproductive concerns and claim the universal importance of their beliefs. As Ross and Solinger write in *Reproductive Justice: An Introduction*, "Reproductive human rights start with the acknowledgement that a person has an inherent human right to control her own body and then seeks to use the political process to express this right and the judicial process to protect this right."[8] Reproductive justice necessitates that both public and private entities respect individual rights to bodily autonomy and highlights the need to respect and support each person's ability to act on those rights. That reproductive justice is needed in the first place illuminates how far the world community still has to go in establishing and enforcing human rights specific to sexuality and reproduction. This is particularly necessary for people lacking power in their societies because reproductive justice "draws attention to the lack of physical, reproductive, and cultural safety for vulnerable people."[9] Human rights theory provides a solid legal and political foundation for reproductive justice.

Second, in addition to human rights, the history of reproductive justice is also intertwined with the African American and Latina/o civil rights movements in the 1960s, some of whose members noted that civil rights include the right to have control over one's own body and the right to choose when and with whom to have children. In those contexts, the ability to have children and to avoid unwanted sterilization spurred activism regarding reproductive justice along the lines of both race and class. The civil rights activist Fannie Lou Hamer, for example, stated famously that "a black woman's body was never hers alone."[10] She also called forced sterilizations of African American women in the Jim Crow South "Mississippi appendectomies," noting that the persistent state violation of women's rights to have children was so common that many considered it a standard medical procedure.[11] Although an end to forced sterilization (especially for incarcerated people) was one issue among many in the African American civil rights movement, the sexism endemic to the movement limited women's ability to gain positions of power and to influence its overall direction. Some civil rights and later black power leaders argued that contraception was a form of black genocide: in other words, any kind of birth control, including abortion, was white supremacists' way of using technology to limit the number of African Americans being born.[12] Unsurprisingly, African American women objected to this characterization of contraception—not because they were unconcerned that many contraceptive advocates had eugenic or population-control aims but because a focus on limiting their reproductive autonomy narrowed analysis of racist oppressions to one when multiple factors were at play.

African American women shaped the US women's rights and women's health movements of the 1970s and 1980s on their own terms, even though their arguments for what would later be called intersectional feminism (feminism that included different axes of oppression, such as race, class, and disability status) would often not be clearly heard, much less acted on, by the white feminist leadership of mainstream organizations like the National Organization for Women (NOW). They supported contraception and safe and legal abortion and at the same time identified forced sterilization of women of color as a central feminist issue. "In so doing," scholars of reproductive activism write, "they negotiated a space that at once distanced them from white feminists who prioritized legal abortion and birth control to the exclusion of other reproductive rights issues and those black Nationalists who declared all contraception and abortion genocidal."[13] As white women tended to face the reverse problem regarding desired sterilization—doctors would often not sterilize them if they had not had three or more children and reached a certain age—they usually did not perceive forced sterilization as a major issue.[14] This lack of attention to sterilization led some African American women to create their own organizations based on a growing interest in intersectional reproductive justice–oriented feminism.

In the mid-1980s, African American groups began to organize in order to manifest their own literature and networks regarding black women's health care. The National Black Women's Health Project (NBWHP) was founded by Byllye Y. Avery in 1983 after a conference at Spelman College, and it published *Body & Soul: The Black Women's Guide to Physical Health and Emotional Well-Being* in 1994. In 1987, Loretta Ross organized the first National Conference on Women of Color and Reproductive Rights at Howard University.[15] In the 1990s, the Committee on Women, Population, and the Environment (CWPE) initiated a campaign to raise awareness about and to challenge Children Requiring a Caring Kommunity (CRACK), later renamed Project Prevention, a privately funded organization founded in 1989 that paid women addicted to drugs $200 to be sterilized or to use long-acting contraceptives.[16] In 1998, the NBWHP (renamed the Black Women's Health Imperative four years later) published another book called *Our Bodies, Our Voices, Our Choices*—echoing the title of the well-known grassroots feminist sexual and reproductive health guide first published in 1970, *Our Bodies, Ourselves*.[17]

Despite the advances of (largely cisgender- and white-led) second-wave feminism, the African American women's health movement, and the civil rights movement in the United States, state-sponsored sterilization of poor women, women of color, imprisoned women, and women with disabilities continued to occur across the US throughout the 1970s. As the law professor Dorothy E. Roberts notes, in 1970, 200,000 sterilization operations were performed in the US, and in 1980, more than 700,000 were performed, a disproportionate number of them on women of color.[18] Women's health activism in the Latina/o community organized around ending state sterilization abuse began in the mid-1970s.

Hundreds of women were sterilized without their knowledge from 1969 to 1973 at the University of Southern California–Los Angeles County Medical Center. In addition to Puerto Rican and Mexican-origin women, other Latinas, poor women, and women of color were sterilized in teaching hospitals across the nation.[19] Ten Latina women filed a federal class action lawsuit against the Los Angeles County Hospital, arguing that they had been sterilized against their will because they did not understand English well enough to agree to the procedures. Although the women lost the case, California hospitals changed their obstetric and gynecological practices to accommodate non-native English speakers, such as printing information sheets in different languages and giving patients under twenty-one years old seventy-two hours to consider their decision so that involuntary sterilizations would not happen again.[20]

In 1970, 200,000 sterilization operations were performed in the US; and in 1980, more than 700,000 were performed, a disproportionate number of them on women of color.

Additionally, a group of women including Helen Rodríguez-Trías founded the Committee to End Sterilization Abuse (CESA) in 1974, which in turn created a coalition of women's health activist groups that developed regulations to protect women patients at public hospitals in New York City.[21] The related Committee for Abortion Rights and against Sterilization Abuse (CARASA) was established in 1977 and published *Women under Attack: Abortion, Sterilization Abuse, and Reproductive Freedom* on their findings in 1979. Both committees were instrumental in establishing US federal guidelines to limit forced sterilization through the Department of Health, Education, and Welfare in 1979, even though the practice continued in some areas of the United States, including California and Tennessee, through the 2010s.[22] As Ross points out, "Women of color, who continuously face strategies of population control through eugenics-based ideologies, must fight equally as hard for the right to have children."[23]

Along with human, civil, and women's right activism, intersectional theories are also critical to understanding reproductive justice. Ross and Solinger state that "reproductive justice is the application of the concept of intersectionality to reproductive politics in order to achieve human rights."[24] The theoretical roots of reproductive justice include black feminist theory, self-help theory, critical race and critical feminist theory, human rights theory, standpoint theory, and womanist ethics and religion theory.[25] Standpoint theory—the idea that knowledge is situated along different axes of power and marginalization—is particularly important, Ross writes, because "as black women, we occupy both an insider and outside position within the feminist movement, the African American community, and in gender-nonconforming spaces."[26] It is no accident that reproductive justice emerged not from academic circles but from activists of color, many of whom suffered from injuries to their reproductive health and human dignity. It is undeniable that "women of color are ideologically leading the movement."[27]

In sum, the experiences of reproductive justice advocates as members of marginalized groups shaped their abilities to identify with systematic injustice and to analyze the mechanisms necessary to manifest political and social change. As the authors of *Radical Reproductive Justice* note, "intersecting forces produce differing reproductive experiences that shape each individual's life. While every human being has the same human rights, our intersectional identities require different considerations to achieve reproductive justice."[28] A foundation in human, civil, and women's rights and also sensitivity to the specific needs of individuals and groups provide reproductive justice activists the intellectual and practical grounding that they need to facilitate the changes that would make the reproductive lives of all healthier and safer.

Reproductive Justice Activism

The individuals and groups involved in the US reproductive justice movement started to work together in the early 1990s. They took inspiration first from global women's health movement events in 1994 (Cairo) and 1995 (Beijing) and second from each other regarding domestic federal health initiatives. Some of the SisterSong Collective for Reproductive Justice "founding mothers" were connected with the Campaign for Women's Health, a national coalition of organizations formed in 1990.[29] In 1992, six of those organizations in turn founded the Women of Color Coalition for Reproductive Health Rights (WOCCRHR), including Asian and Pacific Islanders for Choice, NBWHP, and the Native American Women's Health and Education Resource Center. On the international scale, members of the WOCCRHR participated in the International Conference on Population and Development in Cairo in September 1994, establishing global intellectual and advocacy connections between reproductive health activists and poverty and sexual abuse.[30] Also, the Platform for Action of the fourth United Nations Commission on the Status of Women conference in Beijing in 1995 featured a section on women and health that emphasized sexual and reproductive rights. It suggested specific actions, including woman-centered clinical trials; the regularized provision of safe, effective, and affordable contraceptives; and guaranteed self-determination, equality, and sexual and reproductive security.[31] The Platform for Action provided yet another source of motivation and focused planning for the emerging US reproductive justice community, expanding on the original UDHR.

American reproductive activists found encouragement and direction from each other as well. In June 1994, twelve African American women working in the reproductive health and rights movement established the specific concept of reproductive justice. Gathered for a conference in Chicago sponsored by the Illinois Pro-Choice Alliance and the Ms. Foundation for Women, a group of reproductive justice activists decided to draft a unified response to the Clinton administration's ultimately failed Health Security Act (the universal health-care reform act), first proposed in 1993. Over eight hundred African American women signed an advertisement published in the *Washington Post* and *Roll Call* newspapers in August 1994, decrying the act's lack of attention to preventative health care for women of color, including contraception and abortion.[32]

Writing and publishing that group response energized participants in that conference, and some of its signatories decided to name their new coalition Women of African Descent for Reproductive Justice (WADRJ), renamed SisterSong Collective for Reproductive Justice in 1997 under the leadership of Luz Rodriguez. The coalition included participants from sixteen women-of-color groups active in promoting reproductive health.[33] In fostering collaborative scholarship and action, SisterSong "provided a much-needed space and analysis for these women, allowing them to see connections between themselves as well as the similarities between the past and present reproductive oppressions they faced as a direct result of their identities."[34] They organized to respond to governmental initiatives on women's reproductive health care and to chart and advocate for their own vision of a healthy human society grounded in reproductive justice. That vision reframes the analysis of the history of contraception moving forward.

From Choice to Justice: The Importance of Reproductive Justice to the History of Contraception

As noted above, women-of-color activists first created their own health-oriented organizations based around problems within their specific ethnic groups and also joined forces in SisterSong to advocate for solutions to problems common across communities. They made these decisions when mainstream, largely white-led organizations did not take seriously reproductive health problems that included both structural racism and sexism. Organizations like Planned Parenthood and the Feminist Women's Health Network based much of their thought and action on the concept of choice alone: women needed a full range of choices for fertility-related purposes, including contraception, pregnancy, birth, and infant care. Their leaders were already fighting multifront battles for legalizing new nonhormonal contraceptives like cervical caps; opposing harmful hormonal contraceptives like Depo-Provera; protecting patients and clinic staff from antiabortion protesters; and securing ongoing adequate funding to provide services to poor women. They did not take on systemic racism in reproductive care as well.[35] In sum, "women of color were frustrated with the limitations of the privacy-based pro-choice movement that did not fully incorporate the experiences of women of color, and the failure of the pro-choice movement to understand the impact of white supremacist thinking on the lives of communities of color."[36]

The absence of white feminist leadership for tackling the combined effects of racism and sexism created a gap that women of color would fill with their own intersectional thinking and writing. They criticized the notion that the right to contraception was based in the constitutional right to privacy. This right is not expressed specifically in the original text but put forward in majority US Supreme Court decisions like *Griswold v. Connecticut* (1965), which established the right of married people to possess contraception for their personal use.[37] Women of color saw the limitations in basing arguments for health care in privacy and thus "created a radical shift from 'choice' to 'justice' to locate women's autonomy and self-determination in international human rights standards and laws, rather than in the constitutionally limited concepts of individual rights and privacy."[38] In establishing reproductive justice theory, women of color drew attention to the conceptual and practical inadequacies of deriving a vision for reproductive health and safety from choice alone: "The concept of choice masks the different economic, political, and environmental contexts in which women live their reproductive lives. ... individual choices have only been as capacious and empowering as the resources any woman can turn to in her community."[39] It is critical to acknowledge that "choice" takes on different casts depending on local resources and individual, local, and community reproductive histories.

People make sexual and reproductive health decisions not just with private, individual motives but also in the contexts of their relationships, families, and communities. In developing reproductive justice, women of color turned attention to contextualizing the ability to make informed contraceptive and reproductive decisions as part of advocating for the rights of disadvantaged groups. "Another problem with 'choice,'" Ross and Solinger argue, "is that this market concept strongly refers to the preferences of the individual and suggests that each woman makes her own reproductive choices freely, unimpeded by considerations of family and community. ... Reproduction is a

It is critical to acknowledge that "choice" takes on different casts depending on local resources and individual, local, and community reproductive histories.

biological event and also a social (family and community-based) event, and ... the concept of individual choice cannot capture the context in which persons do or do not become parents."[40] Rights advocacy can be particularly effective in a group that demands attention to racism as well as sexism in pursuit of justice: "Shifting from a focus on individual rights based on privacy, the [reproductive justice] framework invokes collective rights and collective responsibility for organizing our power and acting. ... The [SisterSong] collective's motto [is] 'Doing collectively what we cannot do individually.'"[41]

Among the collective actions of SisterSong and its affiliate organizations, and an example of the need for reproductive justice, is advocacy for safe contraception in underserved communities because "access to effective contraceptive services is crucial to the dignity of women of all races."[42] Reproductive justice advocates are mindful that contraception is just one of the health technologies necessary for people who can become or can make someone else pregnant and who may or may not want to have a child at a given time. They also know the histories of forced sterilization; the use of contraception as a means of population control by governments, NGOs, and private actors; and the ways that government policy, technological and health-care provider availability, and cost may limit contraceptive options in poor and underserved communities. In short, "many women are constrained to make a choice among dangerous or potentially dangerous contraception."[43]

One form of contraception often promoted to poor and undereducated people are long-acting reversible contraceptives (LARCs), usually hormonal-based, which serve as "an effective, if double-edged, solution for persons trying to avoid unintended pregnancy."[44] LARCs may provide protection from pregnancy with limited follow-up, but they also come with health risks, as outlined in chapter 4—and those who take them may not have their full fertility restored for months after their potency ends. Poor people may also be wary of taking part in state-sponsored contraception, given the global history of elites testing new contraceptives on them without full knowledge of the risks and side effects. Moreover, it is clear that "low-income people know that neither LARCs nor any kind of contraceptives can, all by themselves, fix the inequalities—economic, racial, educational, gender, and other—that they face and that may weaken their commitment to contraceptive use, anyway."[45] Thus, contraception is only part of a synthetic approach to reproductive justice based on human rights.

The example of contraception illustrates the intellectual and practical potential of basing advocacy and policymaking on the three main principles of reproductive justice. Where looking at reproductive rights alone would have advocates focused on providing a range of contraceptive options in an underserved community, reproductive justice instead encourages advocates to focus on real-world matters of contraceptive access and at the same time call for changes to public policy that underfunds the program providing that access in the first place. Reproductive justice will be achieved when all people have equal access to safe reproductive health technologies and care and also can raise children in homes and communities that provide related support for healthy development, such as proper nutrition, adequate education, and overall public safety. In the end, "no right can achieve the status of a right if it doesn't apply to all people—and to its corollary: that no right is secure if it is not secure for everybody."[46]

Intersectionality and the Future of Reproductive Justice

Finally, reproductive justice does not imply "that only biologically defined women experience reproductive oppression."[47] In 2006, SisterSong launched its Queer People of Color Caucus (QPOCC) to incorporate into its work the intersection between LGBTQ rights and reproductive justice.[48] Ross and Solinger make this connection explicit: "transgender issues are reproductive justice issues because both domains recognize that the definitions of womanhood, birthing, and mothering (among other concepts involving reproduction) do not fit neatly into the male-female binary."[49] Trans individuals need targeted medical advice when choosing contraceptives, and the reproductive health community in turn requires more education on how best to serve their needs (chapter 6).

In conclusion, inclusivity across multiple identity categories—class, age, income level, ethnic differences within identity categories (multiple possibilities within Asian Pacific Islander communities, for example), and sexual and gender identity—is essential for reproductive justice to come to fruition. These groups face the challenges of balancing similarities of woman-of-color oppressions with differences between and within communities of color. Reproductive justice can be a means of understanding the intersectionality of an individual's or group's sexual and reproductive experiences and also axes of oppression across time and place. It provides a comprehensive vision of what a reproductively just world would look like and tools for evaluating new contraceptives that come on the market. Learning the past and present of reproductive, sexual, and contraceptive history provides a firm foundation for promoting reproductive justice in the future. Some techniques and technologies now in development may indeed forward reproductive justice goals.

THE FUTURE OF CONTRACEPTION

In addition to the contraceptive devices and methods described in chapter 4, new products and ideas are constantly under development. This chapter outlines some of the novel contraceptives that are now undergoing scientific trials and are newly available to people hoping to avoid pregnancy. The first section identifies the problems that many people have with standardized forms of contraception and the ways that the medical and scientific community needs to respond with more nuance to the diversity of bodies needing contraception. It focuses on ongoing developments in scientific knowledge about bodies, particularly those of transgender, overweight, and obese individuals who may not be able to use certain contraceptives safely and effectively. Current contraceptive technologies need to be precisely targeted or modified in order to work for as many people as possible. The second section provides an overview of contraceptive vaccines directed at men and women and the continuing difficulties of creating new contraceptives that meet standards for safety, efficacy, and limited to no side effects. The third section lists a number of new and newly framed possibilities for using technology to reduce or to eliminate the chance of pregnancy, including the Bimek SLV, "outercourse," and sex involving toys, dolls, or robots. The last section revisits the themes articulated in chapter 1 and identifies some of the many threats to reproductive health care in general, not only contraception, that are persistent worldwide. National and local governments, conservative nongovernmental organizations, a lack of information and cooperation between sexual partners, cost, a lack of access to medical care, and on-the-ground distribution problems all hamper access to, use of, and choice of contraception. There is a critical, persistent need for activism and advocacy to support reproductive justice for all.

Contraceptives for Every Body

As scientific knowledge and awareness of variety in human bodies increases, and as medical knowledge in general advances, the more exacting medicines and medical treatments need to become. Those who have health conditions that may interact with (particularly hormonal) contraception must be careful to choose a method that does not conflict with existing medications or increase or cause health problems. This section focuses on two populations requiring specific care and technologies for their contraceptive needs: transgender individuals and overweight and obese women.

Transgender Individuals

All sexually active people who want to prevent pregnancy must consider a method of contraception. However, hormonal treatments for transition may complicate transgender individuals' use of specific technologies and methods, including hormonal birth control pills and IUDs. Transgender teenagers, particularly those who are assigned female at birth (AFAB) and are on a testosterone regimen as part of their transitions, need specialized information and care in order to not become pregnant. Only in the past few years have physicians and health-care professionals started to develop treatment protocols regarding contraceptive technology specifically for transgender individuals—especially teenagers, who are at the most risk for unwanted pregnancy.

For example, inserting IUDs may be complicated for transgender AFAB teenagers because their vaginas may shrink or atrophy due to testosterone therapy.[1] Trans men and nonbinary individuals who are taking testosterone should not use combined estrogen-progesterone

Hormonal treatments for transition may complicate transgender individuals' use of specific technologies and methods, including hormonal birth control pills and IUDs.

pills because the estrogen counteracts testosterone. Furthermore, testosterone therapy alone does not prevent pregnancy. A recent survey of AFAB individuals and their physicians in the United States indicated that they were unaware that testosterone is not a form of contraception on its own and that they needed to use barrier or nonconflicting chemical methods.[2]

At the same time, many transgender men have children already or would like to become pregnant in the present or in the future. The idea that people presenting as men can become pregnant and give birth is often portrayed in popular culture as a subject of mocking or amusement, obscuring these men's need for gynecological and reproductive health care. Furthermore, twenty European countries still retain laws requiring that transgender individuals submit to some form of surgery (genital alterations, chest modifications, and/or the removal of internal organs), some of which result in sterilization, in order to fulfill legal requirements for transition. However, these laws have begun to be struck down across European countries from the late 2000s onward.[3] As one legal scholar put it, "sterilisation should not be a prerequisite for gender recognition."[4] As the European Court of Human Rights continues to scrutinize and strike down sterilization laws in Europe and as transgender rights become more enshrined in international law, more and more people who do not identify as women will be capable of pregnancy in the future. They will need targeted information, technology, and professional care for managing their fertility.

Overweight and Obese Individuals

The same is true for overweight and obese individuals with uteri who do not want to become pregnant or want to space their pregnancies. Contraceptives are among the many medications that are not adequately tested on overweight and obese individuals. Clinical testing of different forms of hormonal contraceptives often excludes study participants with a body mass index (BMI) greater than 25—in other words, those the BMI identifies as overweight or obese.[5] Overweight or obese women using hormonal contraceptives containing estrogen may also be at risk for additional weight gain and at higher risk for blood clots.[6] Overweight or obese women who have health problems sometimes related to weight, such as high blood pressure or diabetes, may have to avoid hormonal pills containing estrogen. Some methods of hormonal contraception (such as IUDs and progestin-only methods) have equal efficacy for those with high body weight, while others (such as the combined pill, the transdermal patch, and the vaginal ring) have less efficacy.[7] The nonhormonal IUD is effective across weight classes, as are barrier methods.

Managing contraception for people choosing surgical weight loss is a challenge as well, and there is little research on postsurgery contraception. Because bariatric surgeries,

Contraceptives are among the many medications that are not adequately tested on overweight and obese individuals.

particularly gastric bypass, affect absorption of nutrition, they likewise affect absorption of medicines like oral contraceptives. Those at risk of pregnancy need to take special care with contraception for one to two years after such major surgery because there is potential for harm to the mother and to the fetus while the body is undergoing rapid weight loss.[8] Oral hormonal pills may not work, so other methods, such as male and female condoms and nonhormonal IUDs, are the best choice. Future research could address the specific contraceptive needs of people undergoing bariatric surgery and the swift weight loss that happens in the months afterward.

Emergency contraception (EC) can be particularly challenging for overweight or obese people. Although progesterone-only pills work well when taken on a daily basis, progesterone-only EC (levonorgestrel, known in the United States as Plan B) may be less effective for people with a BMI over 26 than ulipristal acetate (known in the United States as ella), a selective progesterone receptor modulator.[9] Overweight or obese individuals have the best chance of avoiding unwanted pregnancy if they have a copper IUD inserted within five days of unprotected sex, which may be quite expensive. It is a complicated situation to navigate.

In sum, it is clear that the work of including all bodies in reproductive health care is a matter not only of providing access to certain technologies and methods for users but also of overcoming longstanding prejudices against people with bodies that do not conform to sociocultural standards of gender identity, gender presentation, size, and physical and/or mental ability. Inclusivity can be demonstrated in many ways, such as by using gender-neutral language regarding sexual behavior and pregnancy risk on medical intake forms; having medical professionals use up-to-date terms in conversation with transgender and nonbinary patients; providing gender-neutral restrooms in medical offices; keeping costs low; and continuing to keep pressure on pharmaceutical manufacturers to produce wider ranges of chemical and mechanical contraceptives that accommodate all bodies.[10]

Vaccines

Contraceptive vaccines are part of the medical and scientific community's longstanding attempt to produce contraceptives that have limited side effects, are easily reversible, and can be used by men. All are currently under development in various forms around the world, using either human or animal testing, though finding a magic-bullet contraceptive that meets those criteria remains elusive. Scientists investigating contraceptive vaccines have pursued three avenues of research: an anti-hCG (human chorionic gonadotropin) vaccine for women, antisperm vaccines for men and women, and a hormonal vaccine for men.

First, from the mid-1970s through the mid-1990s, a coalition of nongovernmental organizations involved in human health (including the World Health Organization and the Population Council) conducted research on an anti-hCG vaccine for women. The idea for the vaccine came from scientists in an emerging field called reproductive immunology who examined the ways that the body's own antibodies could prevent conception and embryo implantation. One way to achieve that is by inhibiting the function of hCG, one of the hormones produced by a preimplantation embryo and necessary for a pregnancy to begin. However, it is difficult to block the production of only one factor without interfering with the functioning of others or creating other health problems. Clinical trials in the 1990s revealed that the anti-hCG vaccine interfered with some women's menstrual cycles and that some women were unable to produce enough antibodies for the vaccine to be effective.[11] Research on a new generation of this vaccine took place on women in New Delhi in the 2000s, 25 percent of whom were also unable to produce the necessary antibodies. Although a new combined protein-DNA version of the vaccine underwent testing on animals in 2017, the anti-hCG vaccine is still years away from being approved for the general public.[12]

The second set of vaccines under development fall under the broad category of "immunocontraception," or the triggering of specific antibodies in order to suppress one of the seventy-six conceptive factors required for embryo development and successful embryo implantation to occur.[13] For example, some trial vaccines target sperm-specific proteins. If a vaccine with sperm-specific antigens can produce specific antibodies against these proteins, the proteins can be neutralized, and the physiochemical processes necessary for conception would not take place. Humans diagnosed with infertility produce these antibodies without the vaccine, so the vaccine may trigger the antibodies in people who are otherwise fertile. So far, scientists have tested these vaccines only on mice, and mouse models cannot be translated directly to human models. However, the wide scientific interest in nonhormonal vaccines means that primate and human testing may not be far in the future.[14]

The third vaccine option was designed specifically for men, and the formula was designed with two aims—to reduce sperm production below the threshold for conception and to maintain that contraceptive level for a period of up to fifty-six weeks. The 320 participants in an 2008 to 2012 international study received an injectable contraception vaccine, which included regular doses of both a long-acting progestogen and a long-acting androgen, testosterone undecanoate. They were required to have a normal reproductive system and be in a stable, mutually monogamous relationship for at least one year with a non-pregnant female partner who also had a normal reproductive system. After the vaccine regimen ended, men's sperm counts returned to normal.[15]

The results were striking: the method's effectiveness rate was 92.5 percent, or roughly equivalent to most hormonal methods for women, including the patch, pill, monthly vaginal ring, and shot.[16] However, many of the participants listed mild to moderate related side effects, including pain at the injection site, mood swings, increased libido, depression, acne, and myalgia (muscle pain). An independent Data Safety and Monitoring Committee (DSMC), established by the World Health Organization's Department of Reproductive Health and Research (WHO/RHR) and by Contraceptive Research and Development (CONRAD), terminated the study early due to the men's complaints. As critics of the study point out, these symptoms are also common in approved hormonal birth control for women.[17] This DSMC took these men's complaints regarding these symptoms seriously, but women's complaints about similar or even more severe symptoms are often ignored. As was clear in birth control pill testing in 1950s Puerto Rico and IUD testing in the late 1970s to early 1980s, for example, even severe pain, strokes, permanent physical damage, and death have not been good enough reasons to stop contraceptive technologies for women from being tested and sold. The rapid termination of this study highlights ongoing problems in scientific perceptions of gender difference regarding medical side effects.

Nonetheless, research in the area of nonhormonal and hormonal vaccines continues, despite the significant reported side effects. Antisperm hormonal vaccines delivered in pill and transdermal gel forms have fewer side effects than injections.[18] These vaccines are one potential pathway for the creation of a long-acting reversible male contraceptive method aside from injectables.

Behavior and Barrier Methods

Although behavioral and barrier methods of avoiding pregnancy are findable around the world, contemporary health-care professionals, inventors, toy designers, and roboticists have reframed the ways that people learn, think about, and contextualize them. They have reframed and reintroduced them as ways to fulfill desires for sexual pleasure without the risk of pregnancy.

For example, Planned Parenthood, a popular English-language source for contraceptive information, advice, and services, refers to sexual encounters that do not involve semen or pre-ejaculate entering the vagina as "outercourse." Outercourse as a form of contraception with a partner includes kissing, massage, mutual masturbation, grinding (known in the mid-twentieth century as petting), and discussing erotic fantasies.[19] PP is careful to state that although oral and anal sex will not lead to conception, unless semen or pre-ejaculate gets onto the vulva or into the vagina through related movements, these behaviors can lead to disease transmission without barrier methods of protection. The nonjudgmental tone regarding nonpenetrative sex on PP's website, particularly its attitude toward oral and anal sex, is a far cry from most of the scientific and medical community's silence toward these practices and from the negative attitudes toward them from physicians and advocates in the early twentieth century. PP's advice for outercourse, as well as for other methods, shows that a wider range of sex-positive information is available online to information seekers than there was in print-only sources in the past.

Current inventors are also attracted to the possibilities of designing and manufacturing new technologies that can help people enjoy penile-vaginal sex while avoiding conception. One technology for mechanically obstructing sperm has been under development in Germany. The inventor Clemens Bimek created a device in 2000 that would block sperm without the need for a vasectomy. This device, called the Bimek SLV, works by implanting tiny valves (the size of a gummy bear, according to the product's website) in the vas deferens.[20] After surgery, the valves can be opened and closed with a small switch in the scrotum. When the valve is closed, sperm cells are not released during ejaculation, rendering the seminal fluid sterile. After an initial flurry of news media coverage in the first half of 2016, the project appears to have stalled due to the lack of funding for a factory meeting international manufacturing safety standards and for the organization of human clinical trials (the creators refuse to conduct animal trials). Even if the Bimek SLV never advances beyond its current trial state, more inventors will likely take up the challenge of mechanical sperm-blocking technologies in the future.

The use of technology for sex play is widespread and continues to grow in manifold ways. Any means of internet play can also involve the participants using toys, dolls, and robots themselves or with others. Sex toys can be used alone, with a partner or partners in private space, or in an internet-mediated space with any number of viewers. Nonconceptive sex can involve technologies as the means of participation (computer, tablet, or smartphone) and as implements in creating sexual images themselves.[21]

Beyond the toys that can be found at any online or storefront sex toy shop, inventors have experimented with more humanlike entities that raise concerns for both the dolls and robots themselves and the humans who use them. In the 1990s, the artist Matt McMullen created a silicone female mannequin called a RealDoll, and demand

The use of technology for sex play is widespread and continues to grow in manifold ways.

quickly exceeded supply. As robotic technology has advanced in the last two decades for all kinds of uses, so too has the technology that makes robots seem more human-like. Those advances have been used to create robots specifically designed for sexual use. Since 2017, robot-only and combined robot-human brothels have begun to appear in cities like Toronto, Canada, and Mainz and Dortmund, Germany, where one can purchase a half or full hour with an anatomically correct doll that has changeable costumes and customizable vaginal inlays and hair.[22] One can also order a customizable doll for home use that can move and have a voice app installed that enables a kind of conversation between doll and user.[23]

People engage in sex with toys, dolls, and robots with varying degrees of interactivity for any number of reasons. Building that relationship, however, is fraught with problems. There is a growing debate about whether dolls and robots have rights that need protecting or if robot sex is damaging to people and to the technology itself. People might be having sex with robots to avoid sex with another human as a means of avoiding pregnancy, but they might also be damaging their ability to function in the everyday world in the process—not to mention their ability to treat living people with dignity and respect. On the other hand, some argue that falling in love and having sex with inanimate objects, such as robots, is the logical next step in a world where technology is ever further embedded in everyday human life.[24]

Sex toys, dolls, and robots are not always thought of or used as contraceptives because their use is not limited to those who are capable of conceiving. However, people who purchase a doll or robot for sexual purposes or visit a robot brothel choose consciously to have sex with an entity that cannot become pregnant or make them pregnant. These choices enable sexual pleasure without human interaction or procreation, thus providing an erotic opportunity for those who avoid human-to-human sexual contact.

Access for All

The themes threaded throughout this book continue to resonate in the present and will continue to shape the future of research, manufacture, distribution, and use of contraceptive technologies. The four themes identified in the first chapter as marking the modern history of contraception—power relationships, camouflage technology, method persistence, and a lack of neutrality—continue to shape the reproductive world now.

First, gender and power relationships continue to shape contraceptive use. Although in most countries women now have the legal right to refuse sex with a marital or nonmarital partner, they may consent to sex only reluctantly or under duress to avoid arguments or violence. Gender systems that reinforce male-female inequality and uphold male virility as a virtue cannot change based on the availability of new technologies alone. So women and others with uteri may need to find contraception that an uncooperative partner cannot see or feel in order to protect themselves from undesired pregnancy. Second, it may be less necessary to obtain a contraceptive under the guise of a "camouflage technology" now than in the past, but new legal threats to access, such as conscience clauses for pharmacists in the United States, have emerged. Short- or long-term partners may camouflage their reasons for using one technology over another or their feelings regarding the chance of pregnancy from each other. A different kind of camouflage—known colloquially as "stealthing"—has one partner pretending to use condoms but then removing them without the other's knowledge. The emotions and desires behind individual contraceptive use, misuse, or nonuse remain just as complex in the present as they were in the past.

Third, behavioral methods, such as withdrawal and periodic or complete abstinence, endure in the present, though they are regularly repackaged in new frameworks or with new technological add-ons, such as smartphone apps for timing safe periods without barrier methods. The motivations behind the use of behavioral methods range from adherence to conservative religious principles

The emotions and desires behind individual contraceptive use, misuse, or nonuse remain just as complex in the present as they were in the past.

to a desire for a natural, chemical-free lifestyle and have remarkable persistence. Additionally, IUDs, diaphragms, and cervical caps have been invented and reinvented many times over. Fourth, the availability of contraception and the legal and economic frameworks that build and maintain reproductive health programs are often subject to a country's or state's interest in increasing or limiting population—and the monetary and human resources it is willing to expend. Poorer people and people living in rural areas have fewer opportunities to visit health-care providers than those in urban areas, and the limited health care available to them may be prohibitively expensive. A perfect method of contraception, aside from complete sexual abstinence, is not available, and many contraceptive methods continue to involve short- or long-term side effects or health risks that users deem unacceptable. Reproductive justice is a clear and admirable goal, but it is still a long way from realization.

In the spirit of establishing reproductive justice for all, however, international NGOs, pharmaceutical companies, health professionals, and local activists around the world are working together to improve access, albeit slowly, to contraceptive technologies, information, and services. For example, a partnership of the pharmaceutical company Bayer (which produces hormonal contraceptive pills and three types of IUDs) and NGOs established World Contraception Day in 2007, now marked yearly on September

26, to draw attention to the approximately 225 million women and people with uteri worldwide who currently have unmet needs for contraception and reproductive health care.[25] The WHO published a set of guidelines in 2014, "Ensuring Human Rights in the Provision of Contraceptive Information and Services," based on previous international human rights statements such as the 1994 International Conference on Population and Development's Plan of Action, the 2001 Millennium Development Goals, the United Nations Secretary-General's 2010 Every Woman Every Child initiative (for maternal and child health), and the 2011 creation of the Commission on Information and Accountability for Women's and Children's Health. It also positions reproductive rights as critical to the full exercise of human rights: "The fulfilment of human rights obligations requires that health commodities, including contraceptives, be physically accessible and affordable for all."[26] Despite international political support for access to and information about contraception as a means of improving women's and children's health and longevity and the direct involvement of the pharmaceutical industry, there remain significant obstacles to meeting the needs of those 225 million underserved individuals.

Access to a full range of services, technologies, and information sources regarding contraception and reproductive justice is not improving evenly across the world.

Access to a full range of services, technologies, and information sources regarding contraception and reproductive justice is not improving evenly across the world.

Activists with conservative viewpoints on sexual and reproductive issues are highly organized and mobilized against national and international organizations that promote reproductive justice through the provision of contraception and safe abortion. Often, their objections to contraception are based on strict interpretations of religious beliefs, including the idea that any method of obstructing sperm's path to an egg is a form of abortion.[27] Truly living in a world where reproductive justice is guaranteed for all would require a tectonic shift in how national and international systems structure their investments in reproductive health care—not to mention a massive financial and human resource investment in human health care more broadly. We are far from living in a world where everyone's human rights, including the rights to conceive or not, are respected, and although in some places those rights are firm, they are nonexistent or under attack in others. Supporting access to contraception for all is critical for human health in the present and in the future.

Many people and institutions around the world—nonprofit and for-profit alike—are actively concerned with contraception every day. Pro– and anti–reproductive justice forces clash in print, online, and in person over people's abilities to determine their own sexual and reproductive futures; inventors create and test the newest sex toys and robots; the pharmaceutical and device industries manufacture, test, and ship thousands of condoms and pills per day worldwide; policymakers write white papers and draft recommendations; and health advocates work in communities on behalf of poor and underserved populations. However, the most frequent daily experience of contraception, although shaped by these outside forces, takes place on an individual level. Sex is individually experienced but shaped by a myriad of sociocultural, economic, and political forces outside the individual's control. Greater awareness of the past, present, and future of contraception provides a framework for individual decision making and forwards understanding of the role of contraceptive technology in the making of the human world.

birth control
A material technology, pharmaceutical, chemical, or behavior used to prevent a birth.

camouflage technology
A technology for a sexual, reproductive, or illegal purpose (such as drug use) that is marketed for a legal purpose.

Casti connubii (*On Christian Marriage*) (1930)
The Roman Catholic papal encyclical affirming that periodic or complete abstinence in marriage is the only acceptable form of contraception.

cervical cap
A rubber barrier designed to fit over the cervix and form a physical barrier against sperm. Manufactured in small quantities in England and Germany from the 1920s onward, it had a resurgence of popularity as a result of the US feminist women's health movement in the 1970s and 1980s.

Comstock Act (1873)
The Act of the Suppression of Trade in, and Circulation of, Obscene Literature and Articles of Immoral Use (the Comstock Act). It was the US federal prohibition against the manufacture, marketing, sale, and distribution of goods that could have an indecent or immoral purpose. It was overturned by *United States v. One Package of Japanese Pessaries*, 86 F.2d 737 (1936).

contraception
A material technology, chemical, pharmaceutical, or behavior used to prevent the fertilization of an egg by sperm.

Dalkon Shield
A plastic IUD in the shape of a fishhook with small spikes and dangling strings, developed in the late 1960s by Hugh J. Davis and Irwin Lerner and sold by A. H. Robins as an alternative to the pill. The strings wicked bacteria into the vagina, causing pelvic inflammatory disease, sepsis, infertility, and at least twenty deaths. It was later removed from the US and world markets.

emmenagogue
A pharmaceutical preparation, usually with an herbal base, intended to bring on menstruation.

Enovid
The brand name of the first hormonal pill (10 mg of synthetic progesterone and .15 grams of synthetic estrogen). It was manufactured by the US pharmaceutical company G. D. Searle and was approved by the Food and Drug Administration for contraceptive use in June 1960.

Gräfenberg ring
A round silver IUD with silk strings first designed in the late 1920s by the German gynecologist Ernst Gräfenberg. It was never commercially manufactured.

Humanae vitae (On Human Life) **(1968)**
A Roman Catholic papal encyclical affirming the church's position that only periodic and complete abstinence were acceptable forms of contraception, including natural family planning (NFP). Any other forms of contraception remained sinful.

hormonal pill
A chemical contraceptive first made from a progesterone and estrogen combination and released to the US public in 1957 for menstrual irregularity. It prevents pregnancy by two mechanisms—by thickening mucus to prevent sperm from reaching the egg and by preventing ovulation from taking place. The chemical technology was later reformulated into forms such as the patch, ring, intramuscular shot, and morning-after pill.

intrauterine device (IUD)
A metal or plastic device placed in the uterus by a medical professional. It is designed to prevent sperm from fertilizing an egg by interfering with the passage of sperm.

Mensinga pessary (diaphragm or French cap)
A rubber (later silicone) barrier contraceptive that covers the cervix, adhering to the uterine wall by a watch spring under the rim. It was first created and publicized by the German physician W. P. J. Mensinga in 1882.

natural family planning (NFP)
A form of timing sexual intercourse around ovulation in order to prevent or to improve the chances of pregnancy. Fertile or nonfertile periods are determined with temperature and cervical mucus measurements and charting menstrual periods on a calendar.

reproductive justice
A measurement of ideal reproductive health developed in the late twentieth-century United States that highlights past and present health gaps and inequalities, including access to contraceptive technologies and information. It includes the ability to have a child, the ability to not have a child, and the ability to parent a child in a healthy environment.

spermicide
A chemical combined with a delivery agent designed to kill or to impede the movement of sperm and to place a barrier between sperm and an egg. The most popular spermicide remains nonoxynol-9, first patented in the United States in 1951.

Yuzpe regimen
A method of taking standard birth control pills in a specific combination of ethinyl estradiol and levonorgestrel within seventy-two hours after unprotected sex in order to prevent pregnancy. The method was developed by the Canadian physician A. Albert Yuzpe in 1974 and served as a model for the morning-after pill.

.

NOTES

Chapter 1

1. Norman E. Himes, *Medical History of Contraception* (1936; New York: Schocken Books, 1970), xii.

2. Adele E. Clarke, *Disciplining Reproduction: Modernity, American Life Sciences, and "the Problems of Sex"* (Berkeley: University of California Press, 1998), 8.

3. See, for example, Loretta J. Ross and Rickie Solinger, *Reproductive Justice: An Introduction* (Oakland: University of California Press, 2017).

4. Linda L. Layne, "Introduction," in *Feminist Technology*, ed. Linda L. Layne, Sharra L. Vostral, and Kate Boyer (Urbana: University of Illinois Press, 2010), 29.

5. For examples of this type of behavior and thinking, see Susanne M. Klausen, *Race, Maternity, and the Politics of Birth Control in South Africa, 1910–1939* (Basingstoke, UK: Palgrave Macmillan, 2004), 98, 121–122; Teresa Huhle, *Bevölkerung, Fertilität und Familienplanung in Kolumbien: Eine transnationale Wissensgeschichte im Kalten Krieg* (Bielefeld: Transcript, 2017), 231, 260, 307.

6. Donna J. Drucker, "Astrological Birth Control: Fertility Awareness and the Politics of Non-Hormonal Contraception," accessed April 7, 2019, http://notchesblog.com/2015/06/11/astrological-birth-control-fertility-awareness-and-the-politics-of-non-hormonal-contraception; Natural Cycles, "Quality Assured & Recognised," accessed April 7, 2019; https://www.naturalcycles.com/en/science/certifications; United States Food and Drug Administration, "FDA Allows Marketing of First Direct-to-Consumer App for Contraceptive Use to Prevent Pregnancy," accessed April 7, 2019, https://www.fda.gov/newsevents/newsroom/pressannouncements/ucm616511.htm.

7. Rachel Maines, "Socially Camouflaged Technologies: The Case of the Electromechanical Vibrator," *IEEE Technology and Society Magazine* 8 (June 1989): 3–11.

8. William Green, *Contraceptive Risk: The FDA, Depo-Provera, and the Politics of Experimental Medicine* (New York: New York University Press, 2017).

9. World Health Organization, "Ensuring Human Rights in the Provision of Contraceptive Information and Services: Guidance and Recommendations," accessed April 7, 2019, http://apps.who.int/iris/bitstream/handle/10665/102539/9789241506748_eng.pdf.

Chapter 2

1. Norman E. Himes, *Medical History of Contraception* (1936; New York: Schocken Books, 1970), 20, 107.

2. Himes, *Medical History of Contraception*, 318–321, 187.

3. James Woycke, *Birth Control in Germany, 1871–1933* (London: Routledge, 1988), 40.

4. Janet Farrell Brodie, *Contraception and Abortion in Nineteenth-Century America* (1994; Ithaca, NY: Cornell University Press, 1997), 217; Himes, *Medical History of Contraception*, 321.

5. Andrea Tone, *Devices and Desires: A History of Contraceptives in America* (New York: Hill and Wang, 2002), 120–121; Peter C. Engelman, *A History of the Birth Control Movement in America* (Santa Barbara, CA: Praeger, 2011), 47, 82. Sanger's shifting motivations for contraceptive advocacy over her fifty-year career in the movement are traced in Ellen Chesler, *Woman of Valor: Margaret Sanger and the Birth Control Movement in America* (1992; New York: Simon & Schuster, 2007).

6. Tone, *Devices and Desires*, 126–127; James Reed, *The Birth Control Movement and American Society: From Private Vice to Public Virtue* (1978; Princeton: Princeton University Press, 2014), 114. On the history of birth control clinics in the United States, see Engelman, *History of the Birth Control Movement in America*; Jimmy Elaine Wilkinson Meyer, *Any Friend of the Movement: Networking for Birth Control, 1920–1940* (Columbus: Ohio State University Press, 2004); Cathy Moran Hajo, *Birth Control on Main Street: Organizing Clinics in the United States, 1919–1939* (Urbana: University of Illinois Press, 2010); and Rose Holz, *The Birth Control Clinic in a Marketplace World* (Rochester: University of Rochester Press, 2014).

7. Tone, *Devices and Desires*, 127; Marie Carmichael Stopes, *Contraception (Birth Control): Its Theory, History, and Practice; A Manual for the Medical and Legal Professions* (London: John Bale, Sons and Danielsson, Ltd., 1924), 140–143; Marie Stopes, *The First Five Thousand, Being the First Report of the First Birth Control Clinic in the British Empire* (London: John Bale, Sons and Danielsson, Ltd. 1925), 3–5, 27; Marie Stopes, *Preliminary Notes on Various Technical Aspects of the Control of Contraception* (London: Mothers' Clinic for Constructive Birth Control, 1930), 12–13; Clare Debenham, *Marie Stopes' Sexual Revolution and the Birth Control Movement* (Cham, Switzerland: Palgrave Macmillan, 2018), 93; Diana Wyndham, *Norman Haire and the Study of Sex* (Sydney: University of Sydney Press, 2012), chaps. 4 and 6, Kindle.

8. Atina Grossmann, *Reforming Sex: The German Movement for Birth Control and Abortion Reform, 1920–1950* (New York: Oxford University Press, 1995),

28; Kirsten Leng, *Sexual Politics and Feminist Science: Women Sexologists in Germany, 1900–1933* (Ithaca: Cornell University Press, 2018), 134–135; Sabine Frühstück, *Colonizing Sex: Sexology and Social Control in Modern Japan* (Berkeley: University of California Press, 2003), 116; Susanne M. Klausen, *Race, Maternity, and the Politics of Birth Control in South Africa, 1910–1939* (Basingstoke, UK: Palgrave Macmillan, 2004), 14, 123–125.

9. Sarah Hodges, *Contraception, Colonialism and Commerce: Birth Control in South India, 1920–1940* (2008; Abingdon, UK: Routledge, 2016), 122–123; Sanjam Ahluwalia, *Reproductive Restraints: Birth Control in India, 1877–1947* (Urbana: University of Illinois Press, 2008), 70.

10. Robert L. Dickinson, *Control of Contraception: A Clinical Medical Manual*, 2nd ed. (1931; Baltimore: Williams & Wilkins, 1938), 191–197.

11. Klausen, *Race, Maternity, and the Politics of Birth Control*, 97; Nicole C. Bourbonnais, *Birth Control in the Decolonizing Caribbean: Reproductive Politics and Practice on Four Islands, 1930–1970* (New York: Cambridge University Press, 2016), 151; Ahluwalia, *Reproductive Restraints*, 151; Darshi Thoradeniya, "Birth Control Pill Trials in Sri Lanka: The History and Politics of Women's Reproductive Health (1950–1980)," *Social History of Medicine*, published online first on October 26, 2018, https://doi.org/10.1093/shm/hky076, pp. 9–10.

12. United States v. One Package, 86 F.2d 737 (1936); Hannah M. Stone, "Birth Control Wins," accessed April 7, 2019, https://www.thenation.com/article/birth-control-wins/.

13. Mary McCarthy, *The Group* (1954; New York: Signet, 1963), 72–73. McCarthy called Dottie's diaphragm a "pessary," which is an umbrella term for both contraceptive and noncontraceptive devices used for uterine support.

14. June Levine, *Sisters: The Personal Story of an Irish Feminist* (1982; Cork, Ireland: Attic Press, 2009), chap. 2, Kindle.

15. Charles Knowlton, *Fruits of Philosophy: A Treatise on the Population Question* (1832; Auckland: Floating Press, 2013), chap. 3, Kindle.

16. Himes, *Medical History of Contraception*, 238–245, esp. 243.

17. Dickinson, *Control of Contraception*, 149; Rachel Lynn Palmer and Sarah K. Greenberg, *Facts and Frauds in Woman's Hygiene: A Medical Guide against Misleading Claims and Dangerous Products* (New York: Vanguard Press, 1936), 256.

18. Brodie, *Contraception and Abortion in Nineteenth-Century America*, 72, 74; Dickinson, *Control of Contraception*, 152, 155, 168; Palmer and Greenberg, *Facts and Frauds in Woman's Hygiene*, 132–137, 143, 152–153.

19. Margaret Sanger, *Family Limitation* (n.p., 1914), p. 16, folder 6, box 85, The Margaret Sanger Papers (microfilmed), Sophia Smith Collection, Smith College, Northampton, Massachusetts; Angus McLaren, *A History of Contraception: From Antiquity to the Present Day* (Oxford: Basil Blackwell, 1990), 237, 249 n. 84.

20. Hodges, *Contraception, Colonialism and Commerce*, 124–125.

21. Bourbonnais, *Birth Control in the Decolonizing Caribbean*, 132; Ahluwalia, *Reproductive Restraints*, 68–69.

22. McLaren, *History of Contraception*, 237; Tone, *Devices and Desires*, 157–172; Woycke, *Birth Control in Germany*, 43–45; Hannah M. Stone, *Maternal Health and Contraception: A Study of the Medical Data of Two Thousand Patients from the Maternal Health Center, Newark, NJ* (New York: A. R. Elliott, 1933), 8.

23. Ilana Löwy, "'Sexual Chemistry' before the Pill: Science, Industry and Chemical Contraceptives, 1920–1960," *British Journal for the History of Science* 44 (June 2011): 269, 272; Ettie Rout, *Practical Birth Control: Being a Revised Version of Safe Marriage* (1922; London: William Heinemann (Medical Books) Ltd., 1940), 70.

24. Dickinson, *Control of Contraception*, 149–150; Ahluwalia, *Reproductive Restraints*, 67–68; Bourbonnais, *Birth Control in the Decolonizing Caribbean*, 49, 132; Peeter Tammeveski, "Repression and Incitement: A Critical Demographic, Feminist, and Transnational Analysis of Birth Control in Estonia, 1920–1939," *The History of the Family* 16, no. 1 (2011): 19; Ilana Löwy, "Defusing the Population Bomb in the 1950s: Foam Tablets in India," *Studies in History and Philosophy of Biological and Biomedical Sciences* 43 (September 2012): 583–593.

25. Cecil I. B. Voge, *The Chemistry and Physics of Contraceptives* (London: Jonathan Cape, 1933), 178, 197; Dickinson, *Control of Contraception*, 147, Tone, *Devices and Desires*, 129; Woycke, *Birth Control in Germany*, 115–116.

26. Aquiles J. Sobrero, "Spermicidal Agents: Effectiveness, Use, and Testing," in *Vaginal Contraception: New Developments*, ed. Gerald I. Zatuchni et al. (Hagerstown, MD: Harper & Row, 1979), 48–65; John J. Sciarra, "Vaginal Contraception: Historical Perspective," in *Vaginal Contraception*, 2–12.

27. Bourbonnais, *Birth Control in the Decolonizing Caribbean*, 151, 211; Palmer and Greenberg, *Facts and Frauds in Woman's Hygiene*, 244, 249, 256; Thoradeniya, "Birth Control Pill Trials in Sri Lanka," 10.

28. Brodie, *Contraception and Abortion in Nineteenth-Century America*, 207–209; Himes, *Medical History of Contraception*, 194.

29. Woycke, *Birth Control in Germany*, 38; Klausen, *Race, Maternity, and the Politics of Birth Control in South Africa*, 121; Raúl Necochea López, *A History of Family Planning in Twentieth-Century Peru* (Chapel Hill: University of North Carolina Press, 2014), 83.

30. Tone, *Devices and Desires*, 51, 185–186.

31. Grossmann, *Reforming Sex*, 8, 15; Götz Aly and Michael Sontheimer, *Fromms: How Julius Fromm's Condom Empire Fell to the Nazis*, trans. Shelley Frisch (2007; New York: Other Press, 2009), chap. 3, Kindle; Woycke, *Birth Control in Germany*, 113.

32. Jessica Borge, "'Wanting It Both Ways': The London Rubber Company, the Condom, and the Pill, 1915–1970" (PhD diss., Birkbeck College, University of London, 2017), 71–72, 108, 147–149.

33. Kate Fisher, "Uncertain Aims and Tacit Negotiation: Birth Control Practices in Britain, 1925–1950," *Population and Development Review* 26 (June 2000): 309; Clare Debenham, *Birth Control and the Rights of Women: Post-Suffrage Feminism in the Early Twentieth Century* (London: Tauris, 2014), 72; Claire L. Jones, "Under the Covers? Commerce, Contraceptives, and Consumers in England and Wales, 1880–1960," *Social History of Medicine* 29 (November 2016): 734–756.

34. Woycke, *Birth Control in Germany*, 38; Frühstück, *Colonizing Sex*, 40–41; Aly and Sontheimer, *Fromms*, chap. 6.

35. Himes, *Medical History of Contraception*, 202–206; Tone, *Devices and Desires*, 106, 193–194, 198–199.

36. Joshua Gamson, "Rubber Wars: Struggles over the Condom in the United States," *Journal of the History of Sexuality* 1 (October 1990): 263.

37. Brodie, *Contraception and Abortion in Nineteenth-Century America*, 344 n. 82; Tone, *Devices and Desires*, 31.

38. Ann Dugdale, "Devices and Desires: Constructing the Intrauterine Device, 1908–1988" (PhD diss., University of Wollongong, 1995), 70; Woycke, *Birth Control in Germany*, 114.

39. Dickinson, *Control of Contraception*, 241.

40. Caroline Rusterholz, "Testing the Gräfenberg Ring in Interwar Britain: Norman Haire, Helena Wright, and the Debate over Statistical Evidence, Side Effects, and Intra-uterine Contraception," *Journal of the History of Medicine and Allied Sciences* 72 (October 2017): 463. The rubber made insertion of the IUD smoother but also increased chances of expulsion.

41. Dugdale, "Devices and Desires," 76; Wyndham, *Norman Haire and the Study of Sex*, chap. 7.

42. Rout, *Practical Birth Control*, 58; Dickinson, *Control of Contraception*, 228–229, 240.

43. Frühstück, *Colonizing Sex*, 109, 146–147.

44. Brodie, *Contraception and Abortion in Nineteenth-Century America*, 43; Janet Farrell Brodie, "Menstrual Intervention in the Nineteenth-Century United States," in *Regulating Menstruation: Beliefs, Practices, and Interpretations*, ed. Etienne van de Walle and Elisha P. Renne (Chicago: University of Chicago Press, 2001), 39, 49–50; Palmer and Greenberg, *Facts and Frauds in Woman's Hygiene*, 166–168; Cara Delay, "Pills, Potions, and Purgatives: Women and Abortion Methods in Ireland, 1900–1950," *Women's History Review* 29, no. 3 (2019): 479–499.

45. Felix Freiherr von Oefele, "Anticonceptionelle Arzneistoffe: Ein Beitrag zur Frage des Malthunianismus in alter und neuer Zeit," *Die Heilkunde* 2 (1898): 19 (author's translation).

46. Woycke, *Birth Control in Germany*, 16–18; Rebecca Hodes, "The Culture of Illegal Abortion in South Africa," *Journal of Southern African Studies* 42, no. 1 (2016): 81; Kate Fisher, *Birth Control, Sex, and Marriage in Britain, 1918–1960* (Oxford, UK: Oxford University Press, 2006), 32, 55, 63, 119, 161; Bourbonnais, *Birth Control in the Decolonizing Caribbean*, 140.

47. Sarah Orne Jewett, *The Country of Pointed Firs* (1896; N.p.: CreateSpace Independent Publishing Platform, 2013), chap. 10, Kindle.

48. Kaye Wierzbicki, "A Cup of Pennyroyal Tea," accessed April 7, 2019, http://the-toast.net/2015/05/27/a-cup-of-pennyroyal-tea.

49. *Complete Catalogue of the Products of the Laboratories of Parke, Davis & Co., Manufacturing Chemists, Detroit, Mich., U.S.A.* (Detroit: n.p., 1898), Trade Literature Collection, Smithsonian National Museum of American History, Washington, DC; *Complete Catalog of the Products of the Laboratories of Parke, Davis & Co.* (Detroit: Press of Parke, Davis & Company, 1937), Trade Literature Collection, Smithsonian National Museum of American History, Washington, DC; McLaren, *History of Contraception*, 191; Woycke, *Birth Control in Germany*, 22–23.

50. Himes, *Medical History of Contraception*, 252 n. 46.

51. Stopes, *Contraception (Birth Control)*, 64, Dickinson, *Control of Contraception*, 91; Sølvi Sogner, "Abortion, Birth Control, and Contraception: Fertility Decline in Norway," *Journal of Interdisciplinary History* 34 (Autumn 2003): 225.

52. Frühstück, *Colonizing Sex*, 144–145.

53. Alice B. Stockham, *Tokology: A Book for Every Woman* (New York: R. F. Fenno & Co., 1893); Alice B. Stockham, *Karezza: Ethics of Marriage* (Chicago:

Stockham Publishing Co., 1896); Woycke, *Birth Control in Germany*, 11; Tammeveski, "Repression and Incitement," 25; Stopes, *Contraception (Birth Control)*, 61–62, 56–57; Palmer and Greenberg, *Facts and Frauds in Woman's Hygiene*, 239; Hodges, *Contraception, Colonialism and Commerce*, 125; John Rock and David Loth, *Voluntary Parenthood* (New York: Random House, 1949), 101–102, 150; Kateřina Lišková, *Sexual Liberation, Socialist Style: Communist Czechoslovakia and the Science of Desire, 1945–1989* (Cambridge, UK: Cambridge University Press, 2018), 122, 125–126.

54. Dickinson, *Control of Contraception*, 117; Rout, *Practical Birth Control*, 55; Woycke, *Birth Control in Germany*, 10.

55. Himes, *Medical History of Contraception*, 183; Sogner, "Abortion, Birth Control, and Contraception," 225; Bourbonnais, *Birth Control in the Decolonizing Caribbean*, 139; Kate Fisher, "'She Was Quite Satisfied with the Arrangements I Made': Gender and Birth Control in Britain, 1920–1950," *Past & Present* 169 (November 2000): 169; Fisher, "Uncertain Aims and Tacit Negotiation," 311.

56. Lynn M. Thomas, *Politics of the Womb: Women, Reproduction, and the State of Kenya* (Berkeley: University of California Press, 2003), 34; Lišková, *Sexual Liberation, Socialist Style*, 105; Stopes, *Contraception (Birth Control)*, 72.

57. Lucia Pozzi, "The Problem of Birth Control in the United States under the Papacy of Pius XI," in *Pius XI and America: Proceedings of the Brown University Conference (Providence, February 2010)*, ed. Charles R. Gallagher, David I. Kertzer, and Alberto Meloni (Zurich: Lit Verlag, 2012), 213.

58. Pius XI, *Casti connubii* (*On Christian Marriage*), accessed April 7, 2019, http://w2.vatican.va/content/pius-xi/en/encyclicals/documents/hf_p-xi_enc_19301231_casti-connubii.html; Lucia Pozzi, "The Encyclical *Casti connubii* (1930): The Origin of the Twentieth Century Discourse of the Catholic Church on Family and Sexuality," in *La Sainte Famille: Sexualité, filiation et parentalité dans l'Eglise catholique*, ed. Cécile Vanderpelen-Diagre and Caroline Sägesser (Brussels: Editions de l'Université libre de Bruxelles, 2017), 41–54.

59. Kari Pitkänen, "Contraception in Late Nineteenth- and Early Twentieth-Century Finland," *Journal of Interdisciplinary History* 34 (Autumn 2003): 187; Sogner, "Abortion, Birth Control, and Contraception," 225.

60. Ahluwalia, *Reproductive Restraints*, 46, 71, 78–79; Barbara N. Ramusack, "Embattled Advocates: The Debate over Birth Control in India, 1920–1940," *Journal of Women's History* 1 (Fall 1989): 38, 50, 58.

61. Dickinson, *Control of Contraception*, 258, 265.

62. Tone, *Devices and Desires*, 142–144, 325 n. 79; Dorothy Roberts, *Killing the Black Body: Race, Reproduction, and the Meaning of Liberty* (New York: Vintage, 1998), 66–67; Hajo, *Birth Control on Main Street*, 70–71, 100; Alexandra Minna Stern, "'We Cannot Make a Silk Purse Out of a Sow's Ear': Eugenics in the Hoosier Heartland," *Indiana Magazine of History* 103 (March 2007): 3–38.

63. Woycke, *Birth Control in Germany*, 45–48; Rout, *Practical Birth Control*, 46; Wyndham, *Norman Haire and the Study of Sex*, chap. 9, Kindle.

64. Bent Sigurd Hansen, "Something Rotten in the State of Denmark: Eugenics and the Ascent of the Welfare State," in *Eugenics and the Welfare State: Sterilization Policy in Denmark, Sweden, Norway, and Finland*, ed. Gunnar Broberg and Nils Roll-Hansen (1996; East Lansing: Michigan State University Press, 2005), 38–41.

65. Gunnar Broberg and Mattias Tydén, "Eugenics in Sweden: Efficient Care," in *Eugenics and the Welfare State*, 111; Marjatta Hietala, "From Race Hygiene to Sterilization: The Eugenics Movement in Finland," in *Eugenics and the Welfare State*, 240.

Chapter 3

1. Elaine Tyler May, *America and the Pill: A History of Promise, Peril, and Liberation* (New York: Basic Books, 2010), 57.

2. See, for example, May, *America and the Pill*; Elizabeth Siegel Watkins, *On the Pill: A Social History of Oral Contraceptives, 1950–1970* (Baltimore: Johns Hopkins University Press, 2011); Lara V. Marks, *Sexual Chemistry: A History of the Contraceptive Pill* (2001; New Haven: Yale University Press, 2010); and Jonathan Eig, *The Birth of the Pill: How Four Pioneers Reinvented Sex and Launched a Revolution* (2014; London: Pan Books, 2016).

3. James Reed, *The Birth Control Movement and American Society: From Private Vice to Public Virtue* (1978; Princeton: Princeton University Press, 2014), 337; Peter C. Engelman, *A History of the Birth Control Movement in America* (Santa Barbara, CA: Praeger, 2011), 153. I thank Engelman for clarifying this point in personal communication.

4. Andrea Tone, *Devices and Desires: A History of Contraceptives in America* (New York: Hill and Wang, 2002), 214; May, *America and the Pill*, 24.

5. Laura Briggs, *Reproducing Empire: Race, Sex, Science, and U.S. Imperialism in Puerto Rico* (Berkeley: University of California Press, 2002), 138–139; May, *America and the Pill*, 27–32.

6. Watkins, *On the Pill*, 32; May, *America and the Pill*, 34; Pamela Verma Liao and Janet Dollin, "Half a Century of the Oral Contraceptive Pill: Historical

Review and View to the Future," *Canadian Family Physician* 58 (December 2012): e757. An alternative perspective centers on the role played by a Belgian Catholic physician in testing the lower-dosage pill, Anovlar. This pill was released by the German pharmaceutical company Schering AG (now Bayer) in Europe and Australia seven months after the FDA approved Enovid. See Karl van den Broeck, Dirk Janssens, and Paul Defoort, "A Forgotten Founding Father of the Pill: Ferdinand Peeters, MD," *European Journal of Contraception and Reproductive Health Care* 17, no. 5 (October 2012): 321–328.

7. Beth Bailey, *Sex in the Heartland* (Cambridge, MA: Harvard University Press, 2002), 111.

8. Bailey, *Sex in the Heartland*, 128–129.

9. Barbara Seaman, *The Doctor's Case against the Pill* (New York: Avalon, 1969); May, *America and the Pill*, 130–131.

10. Watkins, *On the Pill*, 108–113; May, *America and the Pill*, 132–133.

11. Sheryl Burt Ruzek, *The Women's Health Movement: Feminist Alternatives to Medical Control* (New York: Praeger, 1978); Wendy Kline, *Bodies of Knowledge: Sexuality, Reproduction, and Women's Health in the Second Wave* (Chicago: University of Chicago Press, 2010); Sandra Morgen, *Into Our Own Hands: The Women's Health Movement in the United States, 1969–1990* (New Brunswick, NJ: Rutgers University Press, 2002); Michelle Murphy, *Seizing the Means of Reproduction: Entanglements of Feminism, Health, and Technoscience* (Durham: Duke University Press, 2012); Jennifer Nelson, *More than Medicine: A History of the Feminist Women's Health Movement* (New York: New York University Press, 2015).

12. Rada Drezgić, "Politics and Practices of Fertility Control under the State Socialism," *History of the Family* 15, no. 2 (2010): 200.

13. Tomáš Sobotka, "The Stealthy Sexual Revolution? Birth Control, Reproduction, and Family under State Socialism in Central and Eastern Europe," in *"Wenn die Chemie Stimmt": Geschlechterbeziehungen und Geburtenkontrolle im Zeitalter der "Pille" / Gender Relations and Birth Control in the Age of the "Pill"*), ed. Lutz Niethammer and Silke Satjukow (Göttingen: Wallstein Verlag, 2016), 140; Agata Ignaciuk, "Reproductive Policies and Women's Birth Control: Practices in State-Socialist Poland (1960s–1980s)," in *"Wenn die Chemie Stimmt,"* 311.

14. Boris Denisov and Victoria Sakevich, "Birth Control in Russia: A Swaying Policy," in *"Wenn die Chemie Stimmt,"* 254–255.

15. Elise Andaya, *Conceiving Cuba: Reproduction, Women, and the State in the Post-Soviet Era* (New Brunswick, NJ: Rutgers University Press, 2014), 84.

16. Kateřina Lišková, *Sexual Liberation, Socialist Style: Communist Czechoslovakia and the Science of Desire, 1945–1989* (Cambridge, UK: Cambridge University Press, 2018), 124; Drezgić, "Politics and Practices of Fertility Control," 200.

17. Rada Drezgić, "Fertility Control and Gender (In)equality under Socialism: The Case of Serbia," in *"Wenn die Chemie Stimmt,"* 276.

18. Amy Kaler, *Running after Pills: Politics, Gender, and Contraception in Colonial Zimbabwe* (Portsmouth, NH: Heinemann, 2003), 160–162, esp. 162.

19. Kaler, *Running after Pills*, 1.

20. Kaler, *Running after Pills*, 22.

21. Takudzwa S. Sayi, "Addressing Limited Contraceptive Options and Inconsistent Use in Zimbabwe," accessed April 7, 2019, https://www.prb.org/contraceptive-use-zimbabwe.

22. Liao and Dollin, "Half a Century of the Oral Contraceptive Pill," E758.

23. On Norplant, which was available in the United States from 1990 to 2000, see Anita Hardon, "Norplant: Conflicting Views on Its Safety and Acceptability," in *Issues in Reproductive Technology: An Anthology*, ed. Helen Bequaert Holmes (New York: Garland, 1992), 11–30; Tone, *Devices and Desires*, 288; and Elizabeth Siegel Watkins, "From Breakthrough to Bust: The Brief Life of Norplant, the Contraceptive Implant," *Journal of Women's History* 22 (Fall 2010): 88–111.

24. William Green, *Contraceptive Risk: The FDA, Depo-Provera, and the Politics of Experimental Medicine* (New York: New York University Press, 2017), chap. 2, Kindle.

25. Kline, *Bodies of Knowledge*, 97–125, esp. 99, 103; Green, *Contraceptive Risk*, chap. 1.

26. Green, *Contraceptive Risk*, chap. 4. For the Indian Health Service's use of Depo-Provera with Native American women, see Barbara Gurr, *Reproductive Justice: The Politics of Health Care for Native American Women* (New Brunswick, NJ: Rutgers University Press, 2014), 126–127.

27. Barbara Klugman, "Balancing Means and Ends: Population Policy in South Africa," *Reproductive Health Matters* 1 (May 1993): 44–57, esp. 52–54.

28. Kate Law, "At Your Service: The Role of the Historian in Contemporary Reproductive Rights Debates," accessed April 7, 2019, https://nursingclio .org/2018/09/13/at-your-service-the-role-of-the-historian-in-contemporary -reproductive-rights-debates; Kate Law, "Fighting Fertility: Depo-Provera, South Africa, and the British Anti-Apartheid Movement," accessed April 7, 2019, https://perceptionsofpregnancy.com/2016/11/28/fighting-fertility -depo-provera-south-africa-and-the-british-anti-apartheid-movement.

29. Rebecca Hodes, "HIV/AIDS in South Africa," accessed April 7, 2019, http://africanhistory.oxfordre.com/view/10.1093/acrefore/9780190277734 .001.0001/acrefore-9780190277734-e-299.

30. Diana Cooper et al., "Coming of Age? Women's Sexual and Reproductive Health after Twenty-One Years of Democracy in South Africa," *Reproductive Health Matters* 24, no. 48 (2016): 80.

31. Cooper et al., "Coming of Age?," 81.

32. For a systematic review of research on this drug interactivity through September 2015, see Kavita Nanda et al., "Drug Interactions between Hormonal Contraceptives and Antiretrovirals," *AIDS* 31, no. 7 (2017): 917–952.

33. L. L. Wynn and Angel M. Foster, "The Birth of a Global Reproductive Health Technology: An Introduction to the Journey of Emergency Contraception," in *Emergency Contraception: The Story of a Global Reproductive Health Technology*, ed. Angel M. Foster and L. L. Wynn (New York: Palgrave Macmillan, 2012), 5.

34. Heather Munro Prescott, *The Morning After: A History of Emergency Contraception in the United States* (New Brunswick, NJ: Rutgers University Press, 2011), 71.

35. Prescott, *The Morning After*, 103–106.

36. Wynn and Foster, "Birth of a Global Reproductive Health Technology," 17.

37. Prescott, *The Morning After*, 123.

38. Nelly Oudshoorn, *The Male Pill: A Biography of a Technology in the Making* (Durham: Duke University Press, 2003), 34–38; Miriam Klemm, "Overshadowed by the Pill–Die Entwicklung männlicher Langzeitverhütungsmittel," *Sexuologie–Zeitschrift fur Sexualmedizin, Sexualtherapie und Sexualwissenschaft* 24, nos. 1–2 (2016): 14.

39. Klemm, "Overshadowed by the Pill," 14.

40. May, *America and the Pill*, 50 (emphasis in original).

Chapter 4

1. Elise Andaya, *Conceiving Cuba: Reproduction, Women, and the State in the Post-Soviet Era* (New Brunswick, NJ: Rutgers University Press, 2014), 42–44.

2. Chikako Takeshita, *The Global Biopolitics of the IUD: How Science Constructs Contraceptive Users and Women's Bodies* (Cambridge, MA: MIT Press, 2012), 13–15, 129–133; Nicole J. Grant, *The Selling of Contraception: The Dalkon Shield Case, Sexuality, and Women's Autonomy* (Columbus: Ohio State University Press, 1992), 38; Ann Dugdale, "Devices and Desires: Constructing the Intrauterine Device, 1908–1988" (PhD diss., University of Wollongong, 1995), 127.

3. Takeshita, *Global Biopolitics of the IUD*, 138; Grant, *Selling of Contraception*, 44; Barbara Seaman, *The Doctor's Case against the Pill* (New York: Avalon, 1969), 1–4.

4. Takeshita, *Global Biopolitics of the IUD*, 166; Grant, *Selling of Contraception*, 24, 51, 67, 176, 200.

5. Takeshita, *Global Biopolitics of the IUD*, 206, 222, 228, 242; Grant, *Selling of Contraception*, 40, 147.

6. Seaman, *Doctor's Case against the Pill*, 204–206.

7. Donna J. Drucker, "Astrological Birth Control: Fertility Awareness and the Politics of Non-Hormonal Contraception," accessed April 7, 2019, http://notchesblog.com/2015/06/11/astrological-birth-control-fertility-awareness-and-the-politics-of-non-hormonal-contraception.

8. Drucker, "Astrological Birth Control"; Dana M. Gallagher and Gary A. Richwald, "Feminism and Regulation Collide: The Food and Drug Administration's Approval of the Cervical Cap," *Women & Health* 15, no. 2 (1989): 87–97; Dana Gallagher, "Cervical Caps and the Women's Health Movement: Feminists as 'Advocate Researchers,'" in *Issues in Reproductive Technology*, ed. Helen Bequaert Holmes (New York: New York University Press, 1992), 87–94; FemCap, "FemCap Improvement History," accessed April 7, 2019, https://femcap.com/about-the-femcap/femcap-improvement-history.

9. Christine Mauck et al., "Lea's Shield: A Study of the Safety and Efficacy of a New Vaginal Barrier Contraception Used with and without Spermicide," *Contraception* 53, no. 6 (1996): 329–335; MedIntim, "Successful History of Development: Caya Contoured Diaphragm," accessed April 7, 2019. https://www.caya.eu/caya-development.

10. Marie Carmichael Stopes, *Contraception (Birth Control): Its Theory, History, and Practice. A Manual for the Medical and Legal Professions* (London: John Bale, Sons & Danielsson, Ltd., 1924), 170.

11. Bikini Condom box, BC: Barrier Methods: Female Condoms pre-1992 folder, box 14, Boston Women's Health Collective Subject Files, 1980–2000, H MS c261, Harvard Medical Library, Francis A. Countway Library of Medicine, Boston, Massachusetts; Department of Health and Human Services, U.S. Food & Drug Administration, "21 CFR Part 884: Obstetrical and Gynecological Devices; Reclassification of Single-Use Female Condom, to Be Renamed Single-Use Internal Condom," *Federal Register* 83, no. 188 (September 27, 2018): 48, 711–48, 713.

12. Joshua Gamson, "Rubber Wars: Struggles over the Condom in the United States," *Journal of the History of Sexuality* 1 (October 1990): 277.

13. Aya Homei, "The Science of Population and Birth Control in Post-War Japan," in *Science, Technology and Medicine in the Japanese Empire*, ed. David G. Wittner and Philip C. Brown (New York: Routledge, 2016), 229, 236; Y. Scott Matsumoto, Akira Koizumi, and Tadahiro Nohara, "Condom Use in Japan," *Studies in Family Planning* 3 (October 1972): 252.

14. Matsumoto, Koizumi, and Nohara, "Condom Use in Japan," 252; Tiana Norgren, *Abortion before Birth Control: The Politics of Reproduction in Postwar Japan* (Princeton: Princeton University Press, 2001), 3, 6; Yasuyo Matsumoto and Shingo Yamabe, "After Ten Years: Has Approval of Oral Contraceptives Really Decreased the Rate of Unintended Pregnancy?," *Contraception* 81 (May 2010): 389–390; Honami Yoshida et al., "Contraception in Japan: Current Trends," *Contraception* 93 (June 2016): 475–477.

15. Emilie Cloatre and Máiréad Enright, "'On the Perimeter of the Lawful': Enduring Illegality in the Irish Family Planning Movement, 1972–1985," *Journal of Law and Society* 44 (December 2017): 472, 473.

16. June Levine, *Sisters: The Personal Story of an Irish Feminist* (1982; Cork, Ireland: Attic Press, 2009), chap. 8, Kindle; Cloatre and Enright, "'On the Perimeter of the Lawful,'" 476.

17. Máiréad Enright and Emilie Cloatre, "Transformative Illegality: How Condoms 'Became Legal' in Ireland, 1991–1993," *Feminist Legal Studies* 26 (November 2018): 261–84; Cloatre and Enright, "'On the Perimeter of the Lawful,'" 481, 484, 489.

18. Julia Vorhölter, "Negotiating Social Change: Ugandan Discourses on Westernization and Neo-Colonialism as Forms of Social Critique," *Journal of Modern African Studies* 50, no. 2 (2012): 305 n. 1.

19. Saurav Jung Thapa, "Uganda Today: Continuing Danger Despite Nullification of Anti-Homosexuality Act," accessed April 7, 2019, https://assets2 .hrc.org/files/assets/resources/Global_Spotlight_Uganda__designed_version __September_25__2015.pdf; Kristen Cheney, "Locating Neocolonialism, 'Tradition,' and Human Rights in Uganda's 'Gay Death Penalty,'" *Africa Studies Review* 55 (September 2012): 78; UNAIDS, "Country Factsheets: Uganda 2017," accessed April 7, 2019, http://www.unaids.org/en/regionscountries/countries/uganda.

20. Robert Poole et al., "Men's Attitude to Condoms and Female-Controlled Means of Protection against HIV and STDs in South-Western Uganda," *Culture, Health, and Sexuality* 2, no. 2 (2000): 197–211; Graham J. Hart et al., "Women's Attitudes to Condoms and Female-Controlled Means of Protection against HIV and STDs in South-Western Uganda," *AIDS Care* 11, no. 6 (1999): 687–698.

21. Makere University, School of Public Health, "Rapid Assessment of Comprehensive Condom Programming in Uganda: Final Report," p. 34, accessed April 7, 2019, http://www.samasha.org/download/Comprehensive-Condom -Programming-Assessment-In-Uganda_Final-Report-October-2015.pdf.

22. Aquiles J. Sobrero, "Evaluation of a New Contraceptive," *Fertility and Sterility* 11, no. 5 (1960): 518–524; Sherwin A. Kaufman, "Simulated Postcoital Test to Determine Immediate Spermicidal Effect of Jelly or Cream Alone," *Fertility and Sterility* 11, no. 2 (1960): 199–209, esp. 204.

23. William H. Masters and Virginia E. Johnson, *Human Sexual Response* (Boston: Little, Brown, 1966); Virginia E. Johnson and William H. Masters, "Intravaginal Contraceptive Study: Phase II. Physiology (a Direct Test for Protective Potential)," *Western Journal of Surgery, Obstetrics, and Gynecology* 71 (May–June 1963): 144–153.

24. Michael J. K. Harper, *Birth Control Technologies: Prospects by the Year 2000* (Austin: University of Texas Press, 1983), 33, 35; Eve W. Paul to Dave Andrews, September 5, 1978, folder 21, box 209, Series II: Classified Files (part 2); Planned Parenthood Federation of America Records, Sophia Smith Collection, Smith College, Northampton, Massachusetts; Christopher Powell and Tim Farley, "Nonoxynol-9 Ineffective in Preventing HIV Infection," accessed April 7, 2019, https://www.who.int/mediacentre/news/notes/release55/en.

25. Andrea Tone, *Devices and Desires: A History of Contraceptives in America* (New York: Hill and Wang, 2002), 285–286.

26. Nelly Oudshoorn, *The Male Pill: A Biography of a Technology in the Making* (Durham: Duke University Press, 2003), 40, 43; Harper, *Birth Control Technologies*, 187–188, 193.

27. Elaine A. Lissner, "Frontiers in Nonhormonal Male Contraceptive Research," in *Issues in Reproductive Technology: An Anthology*, ed. Helen Bequaert Holmes (New York: Garland, 1992), 62–64; Derek Robinson and John Rock, "Intrascrotal Hyperthermia Induced by Scrotal Insulation: Effect on Spermatogenesis," *Obstetrics and Gynecology* 29, no. 2 (February 1967): 217–223; Roger Mieusset et al., "Inhibiting Effect of Artificial Cryptorchidism on Spermatogenesis," *Fertility and Sterility* 43, no. 4 (April 1985): 589–594.

28. Alana Harris, "Introduction: The Summer of '68—Beyond the Secularization Thesis," in *The Schism of '68: Catholicism, Contraception, and 'Humanae Vitae' in Europe, 1945–1975*, ed. Alana Harris (Cham, Switzerland: Palgrave MacMillan, 2018), 5; Paul VI, *Humane vitae (On Human Life)*, accessed April 7, 2019, http://w2.vatican.va/content/paul-vi/en/encyclicals/documents/ hf_p-vi_enc_25071968_humanae-vitae.html.

29. Raúl Necochea López, *A History of Family Planning in Twentieth-Century Peru* (Chapel Hill: University of North Carolina Press, 2014), 136, 141, 146; Teresa Huhle, *Bevölkerung, Fertilität und Familienplanung in Kolumbien: Eine transnationale Wissensgeschichte im Kalten Krieg* (Bielefeld: Transcript, 2017), 246; Lara V. Marks, *Sexual Chemistry: A History of the Contraceptive Pill* (2001; New Haven: Yale University Press, 2010), 231.

30. United States Conference of Catholic Bishops, "NFP Methodology," accessed April 7, 2019, http://www.usccb.org/issues-and-action/marriage-and -family/natural-family-planning/what-is-nfp/methods.cfm; United States Conference of Catholic Bishops, "Benefits of NFP," accessed April 7, 2019, http://www.usccb.org/issues-and-action/marriage-and-family/natural-family -planning/what-is-nfp/benefits.cfm. See also J. J. Billings, *The Ovulation Method: The Achievement or Avoidance of Pregnancy Which Is Reliable and Universally Acceptable* (Melbourne: Advocacy Press, 1983).

31. Toni Weschler, *Taking Charge of Your Fertility: The Definitive Guide to Natural Birth Control, Pregnancy Achievement, and Reproductive Health*, rev. ed. (New York: HarperCollins, 2002), 123; Drucker, "Astrological Birth Control." See also Louise Lacey, *Lunaception: A Feminine Odyssey into Fertility and Contraception* (New York: Coward, McCann, and Geoghegan, 1975), and Arthur Rosenblum, *The Natural Birth Control Book*, 5th ed. (Philadelphia: Aquarian Research Foundation, 1982).

32. Olivia Foster, "Women Are Turning to Birth Control Smartphone Apps for a Reason," accessed April 7, 2019, https://www.theguardian .com/commentisfree/2018/jul/24/women-birth-control-smartphone-apps -contraception-technology; United States Food and Drug Administration, "FDA Allows Marketing of First Direct-to-Consumer App for Contraceptive Use to Prevent Pregnancy," accessed April 7, 2019, https://www.fda.gov/ newsevents/newsroom/pressannouncements/ucm616511.htm; Olivia Sudjic, "'I Felt Colossally Naïve': The Backlash against the Birth Control App," accessed April 7, 2019, https://www.theguardian.com/society/2018/jul/21/ colossally-naive-backlash-birth-control-app.

33. Thomas Scharping, *Birth Control in China, 1949–2000: Population Policy and Demographic Development* (London: Routledge, 2003), 108, 115.

34. Barbara Gurr, *Reproductive Justice: The Politics of Health Care for Native American Women* (New Brunswick, NJ: Rutgers University Press, 2014), 125, 25; Jennifer Nelson, *Women of Color and the Reproductive Rights Movement* (New York: New York University Press, 2003), 140–145; Dorothy Roberts, *Killing the Black Body: Race, Reproduction, and the Meaning of Liberty* (New York: Vintage, 1998), 90.

35. Rebecca Jane Williams, "Storming the Citadels of Poverty: Family Planning under the Emergency in India, 1975–1977," *Journal of Asian Studies* 73 (May 2014): 471–492, esp. 473; Matthew Connelly, "Population Control in India: Prologue to the Emergency Period," *Population and Development Review* 32 (December 2006): 629–667; Davidson R. Gwatkin, "Political Will and Family Planning: The Implications of India's Emergency Experience," *Population and Development Review* 5 (March 1979): 29–59, esp. 29, 33, 38, 47.

36. Karen Hardee et al., "Achieving the Goal of the London Summit on Family Planning by Adhering to Voluntary Rights-Based Family Planning: What Can We Learn from Past Experiences with Coercion?," *International Perspective on Sexual and Reproductive Health* 40 (December 2014): 206–214; Javier Lizarzaburu, "Forced Sterilization Haunts Peruvian Women Decades On," accessed April 7, 2019, https://www.bbc.com/news/world-latin-america-34855804.

37. Loretta J. Ross and Rickie Solinger, *Reproductive Justice: An Introduction* (Oakland: University of California Press, 2017), 52; Rebecca M. Kluchin, *Fit to Be Tied: Sterilization and Reproductive Rights in America* (New Brunswick, NJ: Rutgers University Press, 2009), 22, 54–55, 60, 114–147, esp. 132–137. Hathaway v. Worcester City Hospital, 475 F.2d 701 (1973); Roe v. Wade, 410 U.S. 113 (1973); Doe v. Bolton, 410 U.S. 179 (1973).

38. Huhle, *Bevölkerung, Fertilität und Familienplanung in Kolumbien*, 258–260.

39. Rajesh Varma and Janesh K. Gupta, "Failed Sterilization: Evidence-Based Review and Medico-Legal Ramifications," *BJOG: An International Journal of Obstetrics and Gynecology* 111, no. 12 (December 2004): 1322–1332.

40. Sanket S. Dhruva, Joseph S. Ross, and Aileen M. Gariepy, "Revisiting Essure: Toward Safe and Effective Sterilization," *New England Journal of Medicine* 373, no. 15 (October 8, 2015): e17 (1–3); Essure Permanent Birth Control, "Frequently Asked Questions about Essure," accessed April 7, 2019, http://www.essure.com/faq.

Chapter 5

1. Loretta J. Ross and Rickie Solinger, *Reproductive Justice: An Introduction* (Berkeley: University of California Press, 2017), 56.

2. Jael Silliman, Marlene Gerber Fried, Loretta Ross, and Elena R. Gutiérrez, *Undivided Rights: Women of Color Organize for Reproductive Justice*, 2nd ed. (2004; Chicago: Haymarket Books, 2016), viii.

3. Ross and Solinger, *Reproductive Justice*, 9, 17.

4. Loretta J. Ross, "Conceptualizing Reproductive Justice Theory: A Manifesto for Activism," in *Radical Reproductive Justice: Foundations, Theory, Practice,*

Critique, ed. Loretta Ross et al. (New York: Feminist Press at the City University of New York, 2017), chap. 10, Kindle.

5. Ross and Solinger, *Reproductive Justice*, 68, 69 (emphases in original). See also Asian Communities for Reproductive Justice, "A New Vision for Advancing Our Movement for Reproductive Health, Reproductive Rights, and Reproductive Justice," accessed April 7, 2019, https://forwardtogether.org/tools/a-new-vision.

6. Ross, "Conceptualizing Reproductive Justice Theory," in *Radical Reproductive Justice,* chap. 10; see also Jade S. Sasser, *On Infertile Ground: Population Control and Women's Rights in the Era of Climate Change* (New York: New York University Press, 2018), 144.

7. United Nations, "Universal Declaration of Human Rights," accessed April 7, 2019, http://www.un.org/en/universal-declaration-human-rights.

8. Ross and Solinger, *Reproductive Justice*, 128.

9. Ross, "Conceptualizing Reproductive Justice Theory," in *Radical Reproductive Justice*, chap. 10, Kindle.

10. Loretta J. Ross, "Trust Black Women: Reproductive Justice and Eugenics," in *Radical Reproductive Justice*, chap. 4.

11. Ross, "Trust Black Women"; Sasser, *On Infertile Ground*, 143.

12. Dorothy Roberts, *Killing the Black Body: Race, Reproduction, and the Meaning of Liberty* (New York: Vintage, 1998), 98–102.

13. Silliman et al., *Undivided Rights*, 62.

14. Joyce Wilcox, "The Face of Women's Health: Helen Rodrigues-Trías," *American Journal of Public Health* 92 (April 2002): 566–569, esp. 568.

15. Silliman et al., *Undivided Rights*, 42, 43.

16. Silliman et al., *Undivided Rights*, 17. Project Prevention continues to pay drug-addicted women, often African American, to be sterilized. See Jacquelyn Monroe and Rudolph Alexander, Jr., "C.R.A.C.K: A Progeny of Eugenics and a Forlorn Representation for African Americans," *Journal of African American Studies* 9 (Summer 2005): 19–35.

17. Silliman et al., *Undivided Rights*, 85; *Body & Soul: The Black Women's Guide to Physical Health and Emotional Well-Being*, ed. Linda Villarosa (New York: Perennial, 1994); National Black Women's Health Project, *Our Bodies, Our Voices, Our Choices: A Black Woman's Primer on Reproductive Health and Rights* (Washington, DC: n.p., 1998).

18. Ross and Solinger, *Reproductive Justice*, 51; Roberts, *Killing the Black Body*, 90; Barbara Gurr, *Reproductive Justice: The Politics of Health Care for Native American Women* (New Brunswick, NJ: Rutgers University Press, 2014), 125–126.

19. Silliman et al., *Undivided Rights*, 229; Rebecca M. Kluchin, *Fit to Be Tied: Sterilization and Reproductive Rights in America* (New Brunswick, NJ: Rutgers University Press, 2009), 102–104.

20. Alexandra Minna Stern, "Sterilized in the Name of Public Health: Race, Immigration, and Reproductive Control in Modern California," *American Journal of Public Health* 95 (July 2005): 1128–1138.

21. Silliman et al., *Undivided Rights*, 16, see also 240.

22. Silliman et al., *Undivided Rights*, 39; Committee for Abortion Rights and against Sterilization Abuse, *Women under Attack: Abortion, Sterilization Abuse and Reproductive Freedom* (New York: The Committee, 1979); Wilcox, "The Face of Women's Health," 566–569; Aliya Khan, "Tennessee Judge's 'Birth Control Program' Wasn't 'Controversial': —It Was Coercive," accessed April 7, 2019, https://rewire.news/article/2017/08/02/tennessee-judges -birth-control-program-wasnt-controversial-coercive.

23. Loretta Ross et al., "Introduction," in *Radical Reproductive Justice*.

24. Ross and Solinger, *Reproductive Justice*, 79.

25. Ross, "Conceptualizing Reproductive Justice Theory."

26. Ross, "Conceptualizing Reproductive Justice Theory."

27. Ross et al., "Introduction."

28. Ross et al., "Introduction."

29. Toni M. Bond Leonard, "Laying the Foundations for a Reproductive Justice Movement," in *Radical Reproductive Justice*, chap. 2.

30. Silliman et al., *Undivided Rights*, 48; Leonard, "Laying the Foundations for a Reproductive Justice Movement"; Sasser, *On Infertile Ground*, 143.

31. United Nations, "The United Nations Fourth World Conference on Women—Beijing, China—September 1995; Action for Equality, Development, and Peace," accessed April 7, 2019, http://www.un.org/womenwatch/ daw/beijing/platform/health.htm.

32. Women of African Descent for Reproductive Justice, "Black Women on Universal Health Care Reform," accessed April 7, 2019, https://bwrj .wordpress.com/category/wadrj-on-health-care-reform.

33. Ross, "Conceptualizing Reproductive Justice Theory"; Leonard, "Laying the Foundations for a Reproductive Justice Movement"; Ross et al., "Introduction." The organization is now called SisterSong Women of Color Reproductive Justice Collective.

34. Leonard, "Laying the Foundations for a Reproductive Justice Movement."

35. Donna J. Drucker, "The Cervical Cap in the Feminist Health Movement, 1976–1988," accessed April 7, 2019, http://notchesblog.com/2016/03/24/ the-cervical-cap-in-the-feminist-womens-health-movement-1976-1988.

36. Leonard, "Laying the Foundations for a Reproductive Justice Movement."

37. Griswold v. Connecticut, 381 U.S. 479 (1965). The US Supreme Court did not affirm the right of unmarried Americans to possess contraception until Eisenstadt v. Baird, 405 U.S. 438 (1972).

38. Ross et al., "Introduction."

39. Ross and Solinger, *Reproductive Justice*, 16, 47.

40. Ross and Solinger, *Reproductive Justice*, 102.

41. Leonard, "Laying the Foundations for a Reproductive Justice Movement."

42. Ross and Solinger, *Reproductive Justice*, 152.

43. Ross and Solinger, *Reproductive Justice*, 123.

44. Ross and Solinger, *Reproductive Justice*, 155.

45. Ross and Solinger, *Reproductive Justice*, 156.

46. Ross and Solinger, *Reproductive Justice*, 129.

47. Ross et al., "Introduction."

48. Leonard, "Laying the Foundations for a Reproductive Justice Movement."

49. Ross and Solinger, *Reproductive Justice*, 196.

Chapter 6

1. Linda Prine and Meera Shah, "Long-Acting Reversible Contraception: Difficult Insertions and Removals," *American Family Physician* 98, no. 5 (September 1, 2018): 304–309.

2. Alexis Light, Lin-Fan Wang, Alexander Zeymo, and Veronica Gomez-Lobo, "Family Planning and Contraception Use in Transgender Men," *Contraception* 98 (October 2018): 266–269.

3. Peter Dunne, "Transgender Sterilization Requirements in Europe," *Medical Law Review* 25 (November 2017): 554–581, esp. 556, 560, and 576; Peter Dunne, "YY v Turkey: Infertility as a Pre-Condition for Gender Confirmation Surgery," *Medical Law Review* 23 (December 2015): 646–658.

4. Dunne, "Transgender Sterilization Requirements in Europe," 557.

5. National Heart, Lung, and Blood Institute, "Calculate Your Body Mass Index," accessed April 7, 2019, https://www.nhlbi.nih.gov/health/educational/lose_wt/BMI/bmicalc.htm.

6. Ana Luiza L. Rocha et al., "Safety of Hormonal Contraception for Obese Women," *Expert Opinion on Drug Safety* 16, no. 12 (2017): 1387–1393; Pamela S. Lotke and Bliss Kaneshiro, "Safety and Efficacy of Contraceptive Methods for Obese and Overweight Women," *Obstetrics and Gynecology Clinics of North America* 42 (December 2015): 647–657.

7. Shriya Patel and Lawrence Carey, "Are Hormonal Contraceptives Less Effective in Overweight and Obese Women?," *Journal of the American Academy of Physician Assistants* 31 (January 2018): 11–13.

8. Joël Schlatter, "Oral Contraceptives after Bariatric Surgery," *Obesity Facts* 10, no. 2 (2017): 118–126.

9. Anna Glasier et al., "Can We Identify Women at Risk of Pregnancy despite Using Emergency Contraception? Data from Randomized Trials of Ulipristal Acetate and Levonorgestrel," *Contraception* 84 (October 2011): 363–367.

10. Ronni Hayon, "Gender and Sexual Health: Care of Transgender Patients," *FP Essentials* 449 (October 2016): 27–36; Natalie Ingraham, Erin Wingo, and Sarah C. M. Roberts, "Inclusion of LGBT Persons in Research Related to Pregnancy Risk: A Cognitive Interview Study," *BMJ Sexual and Reproductive Health* 44 (2018): 292–298.

11. Angeline Faye Schrater, "Contraceptive Vaccines: Promises and Problems," in *Issues in Reproductive Technology: An Anthology*, ed. Helen Bequaert Holmes (New York: Garland, 1992), 31–52; Anita Hardon, "Contesting Claims on the Safety and Acceptability of Anti-Fertility Vaccines," *Reproductive Health Matters* 10 (November 1997): 68–81.

12. Gursaran P. Talwar et al., "Current Status of a Unique Vaccine Preventing Pregnancy," *Frontiers in Bioscience, Elite* 9 (June 2017): 321–332.

13. Angela R. Lemons and Rajesh K. Naz, "Birth Control Vaccine Targeting Leukemia Inhibitory Factor," *Molecular Reproduction and Development* 79 (February 2012): 97–106; Angela R. Lemons and Rajesh K. Naz, "Contraceptive Vaccines Targeting Factors Involved in Establishment of Pregnancy," *American Journal of Reproductive Immunology* 66 (July 2011): 13–25.

14. Rajesh K. Naz, "Vaccine for Human Contraception Targeting Sperm Izumo Protein and YLP12 Dodecamer Peptide," *Protein Science* 23 (July 2014): 857–868; see also Martin M. Matzuk et al., "Small-Molecule Inhibition of BRDT for Male Contraception," *Cell* 150 (no. 4, 2012): 673–684; Haruhiko Miyata et al., "Sperm Calcineurin Inhibition Prevents Mouse Fertility with Implications for Male Contraceptive," *Science* 350 (October 23, 2015): 442–445.

15. Hermann M. Behre et al., "Efficacy and Safety of an Injectable Combination Hormonal Contraceptive for Men," *Journal of Clinical Endocrinology and Metabolism* 101 (December 2016): 4779–4788.

16. Planned Parenthood, "Birth Control," accessed April 7, 2019, https://www.plannedparenthood.org/learn/birth-control. The monthly ring was FDA-approved in 2001, and a yearlong ring was approved in 2018. United States Food and Drug Administration, "Approval Package, NuvaRing (Etonogestrel/Ethinyl Estradiol Vaginal Ring)," accessed April 7, 2019, https://www

.accessdata.fda.gov/drugsatfda_docs/nda/2001/21-187_NuvaRing.cfm; and United States Food and Drug Administration, "Drug Approval Package: Annovera (segesterone acetate and ethinyl estradiol)," accessed April 7, 2019, https://www.accessdata.fda.gov/drugsatfda_docs/nda/2018/209627Orig1s 000TOC.cfm.

17. Anna Rhodes, "Yes, Contraceptives Have Side Effects—and It's Time for Men to Put Up with Them Too," accessed April 7, 2019, https://www.independent.co.uk/voices/male-contraceptive-injection-successful-trial-halted-a7384601.html.

18. Niloufar Ilani et al., "A New Combination of Testosterone and Nestorone Transdermal Gels for Male Hormonal Contraception," *Journal of Clinical Endocrinology and Metabolism* 97, no. 10 (2012): 3476–3486; Michael J. Zitzmann et al., "Impact of Various Progestins with or without Transdermal Testosterone on Gonadotropin Levels for Non-Invasive Hormonal Male Contraception: A Randomized Clinical Trial," *Andrology* 5 (May 2017): 516–526.

19. Planned Parenthood, "Abstinence and Outercourse," accessed April 7, 2019, https://www.plannedparenthood.org/learn/birth-control/abstinence-and-outercourse.

20. Bimek SLV, "Contraception You Don't Need to Worry About," accessed April 7, 2019, https://www.bimek.com/this-is-how-the-bimek-slv-works.

21. Timothy Archibald, *Sex Machines: Photographs and Interviews* (Carrboro, NC: Daniel 13 / Process, 2005); Hallie Lieberman, *Buzz: The Stimulating History of the Sex Toy* (New York: Pegasus Books, 2017).

22. Ida Schelenz, "Come on Barbie—Kasteler Bordell testet Sexpuppen," accessed April 7, 2019, https://sensor-magazin.de/come-on-barbie-kasteler-bordell-testet-sexpuppen; Adrian Terhorst, "In Dortmund gibt es das erste Puppen-Bordell Deutschlands," accessed April 7, 2019, https://rp-online.de/nrw/panorama/in-dortmund-gibt-es-das-erste-puppen-bordell-deutschlands_aid-16455925.

23. Breena Kerr, "Future of Sex: How Close Are Robotic Love Dolls?," accessed April 7, 2019, https://www.rollingstone.com/culture/culture-features/future-of-sex-how-close-are-robotic-love-dolls-123749; Allison P. Davis, "Are We Ready for Robot Sex?," accessed April 7, 2019, https://www.thecut.com/2018/05/sex-robots-realbotix.html; Friedemann Karig and Joko Winterscheidt, "Joko zu Besuch im ersten Puppen-Bordell Deutschlands," accessed April 7, 2019, https://www.stern.de/lifestyle/jwd/jwd--magazin--joko-winterscheidt-im-sexpuppen-bordell-in-dortmund-7897394.html.

24. David Levy, *Love and Sex with Robots: The Evolution of Human-Robot Relationships* (New York: Harper Perennial, 2007); John Danaher and Neil

McArthur, eds., *Robot Sex: Social and Ethical Implications* (Cambridge, MA: MIT Press, 2017).

25. Bayer AG, "World Contraception Day: Support Mission #WCD2018," accessed April 7, 2019, https://www.your-life.com/en/for-doctors-parents -etc/about-wcd. The figure of 225 million women with unmet contraceptive needs is from World Health Organization, "Quality of Care in Contraceptive Information and Services, Based on Human Rights Standards: A Checklist for Health Care Providers," accessed April 7, 2019, http://apps.who.int/iris/ bitstream/handle/10665/254826/9789241512091-eng.pdf.

26. World Health Organization, "Ensuring Human Rights in the Provision of Contraceptive Information and Services: Guidance and Recommendations," accessed April 7, 2019, http://apps.who.int/iris/bitstream/handle/ 10665/102539/9789241506748_eng.pdf.

27. For an example of analysis of anti-abortion and anticontraception organizing in the mid-twentieth century United States, see Daniel K. Williams, *Defenders of the Unborn: The Pro-Life Movement before* Roe v. Wade (New York: Oxford University Press, 2016).

BIBLIOGRAPHY

Ahluwalia, Sanjam. *Reproductive Restraints: Birth Control in India, 1877–1947*. Urbana: University of Illinois Press, 2008.

Allbutt, Henry A. *The Wife's Handbook: How a Woman Should Order Herself during Pregnancy, in the Lying-In Room, and after Delivery*. London: Bentley & Co., 1887.

Aly, Götz, and Michael Sontheimer. *Fromms: How Julius Fromm's Condom Empire Fell to the Nazis*. Translated by Shelley Frisch. 2007; New York: Other Press, 2009. Kindle.

Andaya, Elise. *Conceiving Cuba: Reproduction, Women, and the State in the Post-Soviet Era*. New Brunswick, NJ: Rutgers University Press, 2014.

Archibald, Timothy. *Sex Machines: Photographs and Interviews*. Carrboro, NC: Daniel 13, 2005.

Asian Communities for Reproductive Justice. "A New Vision for Advancing Our Movement for Reproductive Health, Reproductive Rights, and Reproductive Justice." Accessed April 7, 2019. https://forwardtogether.org/tools/a-new-vision.

Bailey, Beth. *Sex in the Heartland*. Cambridge, MA: Harvard University Press, 2002.

Bayer AG. "World Contraception Day: Support Mission #WCD2018." Accessed April 7, 2019. https://www.your-life.com/en/for-doctors-parents-etc/about-wcd.

Behre, Hermann M., Michael Zitzmann, Richard A. Anderson, David J. Handelsman, Silvia W. Lestari, Robert I. McLachlan, M. Cristina Meriggiola, et al. "Efficacy and Safety of an Injectable Combination Hormonal Contraceptive for Men." *Journal of Clinical Endocrinology and Metabolism* 101 (December 2016): 4779–4788.

Bikini Condom box. BC: Barrier Methods: Female Condoms pre-1992 folder, box 14, Boston Women's Health Collective Subject Files, 1980–2000, HMS c261, Harvard Medical Library, Francis A. Countway Library of Medicine, Boston, Massachusetts.

Billings, J. J. *The Ovulation Method: The Achievement or Avoidance of Pregnancy Which Is Reliable and Universally Acceptable*. Melbourne: Advocacy Press, 1983.

Bimek SLV. "Contraception You Don't Need to Worry About." Accessed April 7, 2019. https://www.bimek.com/this-is-how-the-bimek-slv-works.

Borge, Jessica. "'Wanting It Both Ways': The London Rubber Company, the Condom, and the Pill, 1915–1970." Ph.D. diss., Birkbeck College, University of London, 2017.

Bourbonnais, Nicole C. *Birth Control in the Decolonizing Caribbean: Reproductive Politics and Practice on Four Islands, 1930–1970*. New York: Cambridge University Press, 2016.

Briggs, Laura. *Reproducing Empire: Race, Sex, Science, and U.S. Imperialism in Puerto Rico*. Berkeley: University of California Press, 2002.

Broberg, Gunnar, and Mattias Tydén. "Eugenics in Sweden: Efficient Care." In *Eugenics and the Welfare State: Sterilization Policy in Denmark, Sweden, Norway, and Finland*, edited by Gunnar Broberg and Nils Roll-Hansen, 77–149. 1996; East Lansing: Michigan State University Press, 2005.

Brodie, Janet Farrell. *Contraception and Abortion in Nineteenth-Century America*. 1994; Ithaca: Cornell University Press, 1997.

Brodie, Janet Farrell. "Menstrual Intervention in the Nineteenth-Century United States." In *Regulating Menstruation: Beliefs, Practices, and Interpretations*, edited by Etienne van de Walle and Elisha P. Renne, 39–63. Chicago: University of Chicago Press, 2001.

Cheney, Kristen. "Locating Neocolonialism, 'Tradition,' and Human Rights in Uganda's 'Gay Death Penalty.'" *Africa Studies Review* 55 (September 2012): 77–95.

Chesler, Ellen. *Woman of Valor: Margaret Sanger and the Birth Control Movement in America*. 1992; New York: Simon & Schuster, 2007.

Clarke, Adele E. *Disciplining Reproduction: Modernity, American Life Sciences, and "the Problems of Sex."* Berkeley: University of California Press, 1998.

Cloatre, Emilie, and Máiréad Enright. "'On the Perimeter of the Lawful': Enduring Illegality in the Irish Family Planning Movement, 1972–1985." *Journal of Law and Society* 44 (December 2017): 471–500.

Committee for Abortion Rights and against Sterilization Abuse. *Women under Attack: Abortion, Sterilization Abuse, and Reproductive Freedom*. New York: The Committee, 1979.

Complete Catalog of the Products of the Laboratories of Parke, Davis & Co. (Detroit: Press of Parke, Davis & Company, 1937), Trade Literature Collection, Smithsonian National Museum of American History, Washington, DC.

Complete Catalogue of the Products of the Laboratories of Parke, Davis & Co., Manufacturing Chemists, Detroit, Mich., U.S.A. (Detroit: n.p., 1898), Trade Literature Collection, Smithsonian National Museum of American History, Washington, DC.

Connelly, Matthew. "Population Control in India: Prologue to the Emergency Period." *Population and Development Review* 32 (December 2006): 629–667.

Cooper, Diana, Jane Harries, Jennifer Moodley, Deborah Constant, Rebecca Hodes, Cathy Mathews, Chelsea Morroni, and Margaret Hoffman. "Coming of Age? Women's Sexual and Reproductive Health after Twenty-One Years of Democracy in South Africa." *Reproductive Health Matters* 24, no. 48 (2016): 79–89.

Danaher, John, and Neil McArthur, eds. *Robot Sex: Social and Ethical Implications*. Cambridge, MA: MIT Press, 2017.

Davis, Allison P. "Are We Ready for Robot Sex?" Accessed April 7, 2019. https://www.thecut.com/2018/05/sex-robots-realbotix.html.

Debenham, Clare. *Birth Control and the Rights of Women: Post-Suffrage Feminism in the Early Twentieth Century*. London: Tauris, 2014.

Debenham, Clare. *Marie Stopes' Sexual Revolution and the Birth Control Movement*. Cham, Switzerland: Palgrave Macmillan, 2018.

Delay, Cara. "Pills, Potions, and Purgatives: Women and Abortion Methods in Ireland, 1900–1950." *Women's History Review* 29, no. 3 (2019): 479–499.

Denisov, Boris, and Victoria Sakevich. "Birth Control in Russia: A Swaying Population Policy." In *"Wenn die Chemie Stimmt": Geschlechterbeziehungen und Geburtenkontrolle im Zeitalter der "Pille"/Gender Relations and Birth Control in the Age of the "Pill,"* edited by Lutz Niethammer and Silke Satjukow, 245–268. Göttingen: Wallstein Verlag, 2016.

Department of Health and Human Services, U.S. Food and Drug Administration. "21 CFR Part 884: Obstetrical and Gynecological Devices; Reclassification of Single-Use Female Condom, to Be Renamed Single-Use Internal Condom." *Federal Register* 83, no. 188 (September 27, 2018): 48,711–48,713.

Dhruva, Sanket S., Joseph J. Ross, and Aileen M. Gariepy. "Revisiting Essure: Toward Safe and Effective Sterilization." *New England Journal of Medicine* 373 (October 8, 2015): e17 (1–3).

Dickinson, Robert L. *Control of Contraception: A Clinical Medical Manual.* 2nd ed. 1931; Baltimore: Williams & Wilkins, 1938.

Drezgić, Rada. "Fertility Control and Gender (In)equality under Socialism: The Case of Serbia." In *"Wenn die Chemie Stimmt": Geschlechterbeziehungen und Geburtenkontrolle im Zeitalter der "Pille"/Gender Relations and Birth Control in the Age of the "Pill,"* edited by Lutz Niethammer and Silke Satjukow, 269–285. Göttingen: Wallstein Verlag, 2016.

Drezgić, Rada. "Politics and Practices of Fertility Control under the State Socialism." *History of the Family* 15, no. 2 (2010): 191–205.

Drucker, Donna J. "Astrological Birth Control: Fertility Awareness and the Politics of Non-Hormonal Contraception." Accessed April 7, 2019. http://notchesblog.com/2015/06/11/astrological-birth-control-fertility-awareness -and-the-politics-of-non-hormonal-contraception.

Drucker, Donna J. "The Cervical Cap in the Feminist Health Movement, 1976–1988." Accessed April 7, 2019. http://notchesblog.com/2016/03/24/ the-cervical-cap-in-the-feminist-womens-health-movement-1976-1988.

Dugdale, Ann. "Devices and Desires: Constructing the Intrauterine Device, 1908–1988." Ph.D. diss., University of Wollongong, 1995.

Dunne, Peter. "Transgender Sterilization Requirements in Europe." *Medical Law Review* 25 (November 2017): 554–581.

Dunne, Peter. "*YY v Turkey:* Infertility as a Pre-Condition for Gender Confirmation Surgery." *Medical Law Review* 23 (December 2015): 646–658.

Eig, Jonathan. *The Birth of the Pill: How Four Pioneers Reinvented Sex and Launched a Revolution.* 2014. London: Pan Books, 2016.

Engelman, Peter C. *A History of the Birth Control Movement in America.* Santa Barbara, CA: Praeger, 2011.

Enright, Máiréad, and Emilie Cloatre. "Transformative Illegality: How Condoms 'Became Legal' in Ireland, 1991–1993." *Feminist Legal Studies* 26 (November 2018): 261–284.

Essure Permanent Birth Control. "Frequently Asked Questions about Essure." Accessed April 7, 2019. http://www.essure.com/faq.

FemCap. "FemCap Improvement History." Accessed April 7, 2019. https://femcap.com/about-the-femcap/femcap-improvement-history.

Fisher, Kate. *Birth Control, Sex, and Marriage in Britain, 1918–1960.* Oxford, UK: Oxford University Press, 2006.

Fisher, Kate. "'She Was Quite Satisfied with the Arrangements I Made': Gender and Birth Control in Britain, 1920–1950." *Past & Present* 169 (November 2000): 161–193.

Fisher, Kate. "Uncertain Aims and Tacit Negotiation: Birth Control Practices in Britain, 1925–1950." *Population and Development Review* 26 (June 2000): 295–317.

Foster, Olivia. "Women Are Turning to Birth Control Smartphone Apps for a Reason." Accessed April 7, 2019. https://www.theguardian.com/commentisfree/2018/jul/24/women-birth-control-smartphone-apps-contraception-technology.

Frühstück, Sabine. *Colonizing Sex: Sexology and Social Control in Modern Japan.* Berkeley: University of California Press, 2003.

Gallagher, Dana. "Cervical Caps and the Women's Health Movement: Feminists as 'Advocate Researchers.'" In *Issues in Reproductive Technology*, edited by Helen Bequaert Holmes, 87–94. New York: New York University Press, 1994.

Gallagher, Dana M., and Gary A. Richwald. "Feminism and Regulation Collide: The Food and Drug Administration's Approval of the Cervical Cap." *Women & Health* 15, no. 2 (1989): 87–97.

Gamson, Joshua. "Rubber Wars: Struggles over the Condom in the United States." *Journal of the History of Sexuality* 1 (October 1990): 262–282.

Glasier, Anna, Sharon T. Cameron, Diana Blithe, Bruno Scherrer, Henri Mathe, Delphine Levy, Erin Gainer, and Andre Ulmann. "Can We Identify Women at Risk of Pregnancy Despite Using Emergency Contraception? Data from

Randomized Trials of Ulipristal Acetate and Levonorgestrel." *Contraception* 84 (October 2011): 363–367.

Grant, Nicole J. *The Selling of Contraception: The Dalkon Shield Case, Sexuality, and Women's Autonomy*. Columbus: Ohio State University Press, 1992.

Green, William. *Contraceptive Risk: The FDA, Depo-Provera, and the Politics of Experimental Medicine*. New York: New York University Press, 2017.

Grossmann, Atina. *Reforming Sex: The German Movement for Birth Control and Abortion Reform, 1920–1950*. New York: Oxford University Press, 1995.

Gurr, Barbara. *Reproductive Justice: The Politics of Health Care for Native American Women*. New Brunswick, NJ: Rutgers University Press, 2014.

Gwatkin, Davidson R. "Political Will and Family Planning: The Implications of India's Emergency Experience." *Population and Development Review* 5 (March 1979): 29–59.

Hajo, Cathy Moran. *Birth Control on Main Street: Organizing Clinics in the United States, 1919–1939*. Urbana: University of Illinois Press, 2010.

Hansen, Bent Sigurd. "Something Rotten in the State of Denmark: Eugenics and the Ascent of the Welfare State." In *Eugenics and the Welfare State: Sterilization Policy in Denmark, Sweden, Norway, and Finland*, edited by Gunnar Broberg and Nils Roll-Hansen, 9–76. 1996; East Lansing: Michigan State University Press, 2005.

Hardee, Karen, Shannon Harris, Marida Rodriguez, Jan Kumar, Lynn Bakamjian, Karen Newman, and Wyn Brown. "Achieving the Goal of the London Summit on Family Planning by Adhering to Voluntary Rights-Based Family Planning: What Can We Learn from Past Experiences with Coercion?" *International Perspective on Sexual and Reproductive Health* 40 (December 2014): 206–214.

Hardon, Anita. "Contesting Claims on the Safety and Acceptability of Anti-Fertility Vaccines." *Reproductive Health Matters* 10 (November 1997): 68–81.

Hardon, Anita. "Norplant: Conflicting Views on Its Safety and Acceptability." In *Issues in Reproductive Technology: An Anthology*, edited by Helen Bequaert Holmes, 11–30. New York: Garland, 1992.

Harper, Michael J. K. *Birth Control Technologies: Prospects by the Year 2000*. Austin: University of Texas Press, 1983.

Harris, Alana. "Introduction: The Summer of '68—Beyond the Secularization Thesis." In *The Schism of '68: Catholicism, Contraception, and 'Humanae Vitae' in Europe, 1945–1975*, edited by Alana Harris, 1–20. Cham, Switzerland: Palgrave MacMillan, 2018.

Hart, Graham J., Robert Pool, Gillian Green, Susan Harrison, Stella Nyanzi, and J. A. G. Whitworth. "Women's Attitudes to Condoms and Female-Controlled Means of Protection against HIV and STDs in South-Western Uganda." *AIDS Care* 11, no. 6 (1999): 687–698.

Hayon, Ronni. "Gender and Sexual Health: Care of Transgender Patients." *FP Essentials* 449 (October 2016): 27–36.

Hietala, Marjatta. "From Race Hygiene to Sterilization: The Eugenics Movement in Finland." In *Eugenics and the Welfare State: Sterilization Policy in Denmark, Sweden, Norway, and Finland*, edited by Gunnar Broberg and Nils Roll-Hansen, 195–258. 1996; East Lansing: Michigan State University Press, 2005.

Himes, Norman E. *Medical History of Contraception*. 1936; New York: Schocken Books, 1970.

Hodes, Rebecca. "The Culture of Illegal Abortion in South Africa." *Journal of Southern African Studies* 42, no. 1 (2016): 79–93.

Hodes, Rebecca. "HIV/AIDS in South Africa." Accessed April 7, 2019. http://africanhistory.oxfordre.com/view/10.1093/acrefore/9780190277734.001.0001/acrefore-9780190277734-e-299.

Hodges, Sarah. *Contraception, Colonialism and Commerce: Birth Control in South India, 1920–1940*. 2008; Abingdon, UK: Routledge, 2016.

Holz, Rose. *The Birth Control Clinic in a Marketplace World*. Rochester: University of Rochester Press, 2014.

Homei, Aya. "The Science of Population and Birth Control in Post-War Japan." In *Science, Technology and Medicine in the Japanese Empire*, edited by David G. Wittner and Philip C. Brown, 227–243. New York: Routledge, 2016.

Huhle, Teresa. *Bevölkerung, Fertilität und Familienplanung in Kolumbien: Eine transnationale Wissensgeschichte im Kalten Krieg*. Bielefeld: Transcript, 2017.

Ignaciuk, Agata. "Reproductive Policies and Women's Birth Control: Practices in State-Socialist Poland (1960s–1980s)." In *"Wenn die Chemie Stimmt"*:

Geschlechterbeziehungen und Geburtenkontrolle im Zeitalter der "Pille"/Gender Relations and Birth Control in the Age of the "Pill," edited by Lutz Niethammer and Silke Satjukow, 305–330. Göttingen: Wallstein Verlag, 2016.

Ilani, Niloufar, Mara Y. Roth, John K. Amory, Ronald S. Swerdloff, Clint Dart, Stephanie T. Page, William J. Bremner, et al. "A New Combination of Testosterone and Nestorone Transdermal Gels for Male Hormonal Contraception." *Journal of Clinical Endocrinology and Metabolism* 97, no. 10 (2012): 3476–3486.

Ingraham, Natalie, Erin Wingo, and Sarah C. M. Roberts. "Inclusion of LGBT Persons in Research Related to Pregnancy Risk: A Cognitive Interview Study." *BMJ Sexual and Reproductive Health* 44 (2018): 292–298.

Jewett, Sarah Orne. *The Country of Pointed Firs.* 1896; N.p.: CreateSpace Independent Publishing Platform, 2013. Kindle.

Johnson, Virginia E., and William H. Masters. "Intravaginal Contraceptive Study: Phase II. Physiology (A Direct Test for Protective Potential)." *Western Journal of Surgery, Obstetrics, and Gynecology* 71 (May–June 1963): 144–153.

Jones, Claire L. "Under the Covers? Commerce, Contraceptives, and Consumers in England and Wales, 1880–1960." *Social History of Medicine* 29 (November 2016): 734–756.

Kaler, Amy. *Running after Pills: Politics, Gender, and Contraception in Colonial Zimbabwe.* Portsmouth, NH: Heinemann, 2003.

Karig, Friedemann, and Joko Winterscheidt. "Joko zu Besuch im ersten Puppen-Bordell Deutschlands." Accessed April 7, 2019. https://www.stern.de/lifestyle/jwd/jwd--magazin--joko-winterscheidt-im-sexpuppen-bordell-in-dortmund-7897394.html.

Kaufman, Sherwin A. "Simulated Postcoital Test to Determine Immediate Spermicidal Effect of Jelly or Cream Alone." *Fertility and Sterility* 11, no. 2 (1960): 199–209.

Kerr, Breena. "Future of Sex: How Close Are Robotic Love Dolls?" Accessed April 7, 2019. https://www.rollingstone.com/culture/culture-features/future-of-sex-how-close-are-robotic-love-dolls-123749.

Khan, Aliya. "Tennessee Judge's 'Birth Control Program' Wasn't 'Controversial': It Was Coercive." Accessed April 7, 2019. https://rewire.news/article/2017/08/02/tennessee-judges-birth-control-program-wasnt-controversial-coercive.

Klausen, Susanne M. *Race, Maternity, and the Politics of Birth Control in South Africa, 1910–1939*. Basingstoke, UK: Palgrave Macmillan, 2004.

Klemm, Miriam. "Overshadowed by the Pill–Die Entwicklung männlicher Langzeitverhütungsmittel." *Sexuologie–Zeitschrift für Sexualmedizin, Sexualtherapie und Sexualwissenschaft* 24, nos. 1–2 (2016): 11–18.

Kline, Wendy. *Bodies of Knowledge: Sexuality, Reproduction, and Women's Health in the Second Wave*. Chicago: University of Chicago Press, 2010.

Kluchin, Rebecca M. *Fit to Be Tied: Sterilization and Reproductive Rights in America*. New Brunswick, NJ: Rutgers University Press, 2009.

Klugman, Barbara. "Balancing Means and Ends: Population Policy in South Africa." *Reproductive Health Matters* 1 (May 1993): 44–57.

Knowlton, Charles. *Fruits of Philosophy: A Treatise on the Population Question*. 1832; Auckland: Floating Press, 2013. Kindle.

Lacey, Louise. *Lunaception: A Feminine Odyssey into Fertility and Contraception*. New York: Coward, McCann, and Geoghegan, 1975.

Law, Kate. "At Your Service: The Role of the Historian in Contemporary Reproductive Rights Debates." Accessed April 7, 2019. https://nursingclio .org/2018/09/13/at-your-service-the-role-of-the-historian-in-contemporary -reproductive-rights-debates.

Law, Kate. "Fighting Fertility: Depo-Provera, South Africa, and the British Anti-Apartheid Movement." Accessed April 7, 2019. https://perceptionsof pregnancy.com/2016/11/28/fighting-fertility-depo-provera-south-africa -and-the-british-anti-apartheid-movement.

Layne, Linda L. "Introduction." In *Feminist Technology*, edited by Linda L. Layne, Sharra L. Vostral, and Kate Boyer, 1–35. Urbana: University of Illinois Press, 2010.

Lemons, Angela R., and Rajesh K. Naz. "Birth Control Vaccine Targeting Leukemia Inhibitory Factor." *Molecular Reproduction and Development* 79 (February 2012): 97–106.

Lemons, Angela R., and Rajesh K. Naz. "Contraceptive Vaccines Targeting Factors Involved in Establishment of Pregnancy." *American Journal of Reproductive Immunology* 66 (July 2011): 13–25.

Leng, Kirsten. *Sexual Politics and Feminist Science: Women Sexologists in Germany, 1900–1933*. Ithaca: Cornell University Press, 2018.

Leonard, Toni M. Bond. "Laying the Foundations for a Reproductive Justice Movement." In *Radical Reproductive Justice: Foundations, Theory, Practice, Critique*, edited by Loretta J. Ross, Lynn Roberts, Erika Derkas, Whitney Peoples, and Pamela Bridgewater, 39–49. New York: Feminist Press at the City University of New York, 2017. Kindle.

Levine, June. *Sisters: The Personal Story of an Irish Feminist*. 1982; Cork, Ireland: Attic Press, 2009. Kindle.

Levy, David. *Love and Sex with Robots: The Evolution of Human-Robot Relationships*. New York: Harper Perennial, 2007.

Liao, Pamela Verma, and Janet Dollin. "Half a Century of the Oral Contraceptive Pill: Historical Review and View to the Future." *Canadian Family Physician* 58 (December 2012): e757–e760.

Lieberman, Hallie. *Buzz: The Stimulating History of the Sex Toy*. New York: Pegasus Books, 2017.

Light, Alexis, Lin-Fan Wang, Alexander Zeymo, and Veronica Gomez-Lobo. "Family Planning and Contraception Use in Transgender Men." *Contraception* 98 (October 2018): 266–269.

Lišková, Kateřina. *Sexual Liberation, Socialist Style: Communist Czechoslovakia and the Science of Desire, 1945–1989*. Cambridge, UK: Cambridge University Press, 2018.

Lissner, Elaine A. "Frontiers in Nonhormonal Male Contraceptive Research." In *Issues in Reproductive Technology: An Anthology*, edited by Helen Bequaert Holmes, 53–69. New York: Garland, 1992.

Lizarzaburu, Javier. "Forced Sterilization Haunts Peruvian Women Decades On." Accessed April 7, 2019. https://www.bbc.com/news/world-latin-america-34855804.

López, Raúl Necochea. *A History of Family Planning in Twentieth-Century Peru*. Chapel Hill: University of North Carolina Press, 2014.

Lotke, Pamela S., and Bliss Kaneshiro. "Safety and Efficacy of Contraceptive Methods for Obese and Overweight Women." *Obstetrics and Gynecology Clinics of North America* 42 (December 2015): 647–657.

Löwy, Ilana. "Defusing the Population Bomb in the 1950s: Foam Tablets in India." *Studies in History and Philosophy of Biological and Biomedical Sciences* 43 (September 2012): 583–593.

Löwy, Ilana. "'Sexual Chemistry' Before the Pill: Science, Industry and Chemical Contraceptives, 1920–1960." *British Journal for the History of Science* 44 (June 2011): 245–274.

Maines, Rachel. "Socially Camouflaged Technologies: The Case of the Electromechanical Vibrator." *IEEE Technology and Society Magazine* 8 (June 1989): 3–11.

Makere University, School of Public Health. "Rapid Assessment of Comprehensive Condom Programming in Uganda: Final Report." Accessed April 7, 2019. http://www.samasha.org/download/Comprehensive-Condom-Programming -Assessment-In-Uganda_Final-Report-October-2015.pdf.

Marks, Lara V. *Sexual Chemistry: A History of the Contraceptive Pill*. 2001; New Haven: Yale University Press, 2010.

Masters, William H., and Virginia E. Johnson. *Human Sexual Response*. Boston: Little, Brown, 1966.

Matsumoto, Yasuyo, and Shingo Yamabe. "After Ten Years: Has Approval of Oral Contraceptives Really Decreased the Rate of Unintended Pregnancy?" *Contraception* 81 (May 2010): 389–390.

Matsumoto, Y. Scott, Akira Koizumi, and Tadahiro Nohara. "Condom Use in Japan." *Studies in Family Planning* 3 (October 1972): 251–255.

Matzuk, Martin M., Michael R. McKeown, Panagis Filippakopoulos, Qinglei Li, Lang Ma, Julio E. Agno, Madeleine E. Lemieux, et al. "Small-Molecule Inhibition of BRDT for Male Contraception." *Cell* 150, no. 4 (2012): 673–684.

Mauck, Christine, Lucinda H. Glover, Eric Miller, Susan Allen, David F. Archer, Paul Blumenthal, Bruce A. Rosenzweig, et al. "Lea's Shield: A Study of the Safety and Efficacy of a New Vaginal Barrier Contraception Used with and without Spermicide." *Contraception* 53, no. 6 (1996): 329–335.

May, Elaine Tyler. *America and the Pill: A History of Promise, Peril, and Liberation*. New York: Basic Books, 2010.

McCarthy, Mary. *The Group*. 1954; New York: Signet, 1963.

McLaren, Angus. *A History of Contraception: From Antiquity to the Present Day*. Oxford: Basil Blackwell, 1990.

MedIntim. "Successful History of Development: Caya Contoured Diaphragm." Accessed April 7, 2019. https://www.caya.eu/caya-development.

Meyer, Jimmy Elaine Wilkinson. *Any Friend of the Movement: Networking for Birth Control, 1920–1940*. Columbus: Ohio State University Press, 2004.

Mieusset, Roger, Helene Grandjean, Arlette Mansat, and Francis Pontonnier. "Inhibiting Effect of Artificial Cryptorchidism on Spermatogenesis." *Fertility and Sterility* 43, no. 4 (April 1985): 589–594.

Miyata, Haruhiko, Yuhkoh Satouh, Daisuke Mashiko, Masanaga Muto, Kaori Nozawa, Kogiku Shiba, Yoshitaka Fujihara, Ayako Isotani, Kazuo Inaba, and Masahito Ikawa. "Sperm Calcineurin Inhibition Prevents Mouse Fertility with Implications for Male Contraceptive." *Science* 350 (October 23, 2015): 442–445.

Monroe, Jacquelyn, and Rudolph Alexander, Jr. "C.R.A.C.K: A Progeny of Eugenics and a Forlorn Representation for African Americans." *Journal of African American Studies* 9 (Summer 2005): 19–35.

Morgen, Sandra. *Into Our Own Hands: The Women's Health Movement in the United States, 1969–1990*. New Brunswick, NJ: Rutgers University Press, 2002.

Murphy, Michelle. *Seizing the Means of Reproduction: Entanglements of Feminism, Health, and Technoscience*. Durham: Duke University Press, 2012.

Nanda, Kavita, Gretchen S. Stuart, Jennifer Robinson, Andrew L. Gray, Naomi K. Tepper, and Mary E. Gaffield. "Drug Interactions between Hormonal Contraceptives and Antiretrovirals." *AIDS* 31, no. 7 (2017): 917–952.

National Black Women's Health Project. *Our Bodies, Our Voices, Our Choices: A Black Woman's Primer on Reproductive Health and Rights*. Washington, DC: n.p., 1998.

National Heart, Lung, and Blood Institute. "Calculate Your Body Mass Index." Accessed April 7, 2019. https://www.nhlbi.nih.gov/health/educational/lose _wt/BMI/bmicalc.htm.

Natural Cycles. "Quality Assured & Recognised." Accessed April 7, 2019. https://www.naturalcycles.com/en/science/certifications.

Naz, Rajesh K. "Vaccine for Human Contraception Targeting Sperm Izumo Protein and YLP12 Dodecamer Peptide." *Protein Science* 23 (July 2014): 857–868.

Nelson, Jennifer. *More than Medicine: A History of the Feminist Women's Health Movement*. New York: New York University Press, 2015.

Nelson, Jennifer. *Women of Color and the Reproductive Rights Movement*. New York: New York University Press, 2003.

Norgren, Tiana. *Abortion before Birth Control: The Politics of Reproduction in Postwar Japan*. Princeton: Princeton University Press, 2001.

Oudshoorn, Nelly. *The Male Pill: A Biography of a Technology in the Making*. Durham: Duke University Press, 2003.

Palmer, Rachel Lynn, and Sarah K. Greenberg. *Facts and Frauds in Woman's Hygiene: A Medical Guide against Misleading Claims and Dangerous Products*. New York: Vanguard Press, 1936.

Patel, Shriya, and Lawrence Carey. "Are Hormonal Contraceptives Less Effective in Overweight and Obese Women?" *Journal of the American Academy of Physician Assistants* 31 (January 2018): 11–13.

Paul, Eve W., to Dave Andrews, September 5, 1978. Folder 21, box 209, Series II: Classified Files (part 2); Planned Parenthood Federation of America Records, Sophia Smith Collection, Smith College, Northampton, Massachusetts.

Paul VI. *Humane vitae* [*On Human Life*]. Accessed April 7, 2019. http://w2.vatican.va/content/paul-vi/en/encyclicals/documents/hf_p-vi_enc_25071968_humanae-vitae.html.

Pitkänen, Kari. "Contraception in Late Nineteenth- and Early Twentieth-Century Finland." *Journal of Interdisciplinary History* 34 (Autumn 2003): 187–207.

Pius XI. *Casti connubii* [*On Christian Marriage*]. Accessed April 7, 2019. http://w2.vatican.va/content/pius-xi/en/encyclicals/documents/hf_p-xi_enc_19301231_casti-connubii.html.

Planned Parenthood. "Abstinence and Outercourse." Accessed April 7, 2019. https://www.plannedparenthood.org/learn/birth-control/abstinence-and-outercourse.

Planned Parenthood. "Birth Control." Accessed April 7, 2019. https://www.plannedparenthood.org/learn/birth-control.

Poole, Robert, Graham Hart, Gillian Green, Susan Harrison, Stella Nyanzi, and Jimmy Whitworth. "Men's Attitude to Condoms and Female-Controlled Means of Protection against HIV and STDs in South-Western Uganda." *Culture, Health, and Sexuality* 2, (no. 2, (2000): 197–211.

Powell, Christopher, and Tim Farley. "Nonoxynol-9 Ineffective in Preventing HIV Infection." Accessed April 7, 2019. https://www.who.int/mediacentre/news/notes/release55/en.

Pozzi, Lucia. "The Encyclical *Casti connubii* (1930): The Origin of the Twentieth Century Discourse of the Catholic Church on Family and Sexuality." In *La Sainte Famille: Sexualité, filiation et parentalité dans l'Eglise catholique*, edited by Cécile Vanderpelen-Diagre and Caroline Sägesser, 41–54. Brussels: Editions de l'Université libre de Bruxelles, 2017.

Pozzi, Lucia. "The Problem of Birth Control in the United States under the Papacy of Pius XI." In *Pius XI and America: Proceedings of the Brown University Conference (Providence, February 2010)*, edited by Charles R. Gallagher, David I. Kertzer, and Alberto Meloni, 209–232. Zurich: Lit Verlag, 2012.

Prescott, Heather Munro. *The Morning After: A History of Emergency Contraception in the United States.* New Brunswick, NJ: Rutgers University Press, 2011.

Prine, Linda, and Meera Shah. "Long-Acting Reversible Contraception: Difficult Insertions and Removals." *American Family Physician* 98, no. 5 (September 1, 2018): 304–309.

Ramusack, Barbara N. "Embattled Advocates: The Debate over Birth Control in India, 1920–1940." *Journal of Women's History* 1 (Fall 1989): 34–64.

Reed, James. *The Birth Control Movement and American Society: From Private Vice to Public Virtue.* 1978; Princeton: Princeton University Press, 2014.

Rhodes, Anna. "Yes, Contraceptives Have Side Effects—and It's Time for Men to Put Up with Them Too." Accessed April 7, 2019. https://www.independent.co.uk/voices/male-contraceptive-injection-successful-trial-halted-a7384601.html.

Roberts, Dorothy. *Killing the Black Body: Race, Reproduction, and the Meaning of Liberty.* New York: Vintage, 1998.

Robinson, Derek, and John Rock. "Intrascrotal Hyperthermia Induced by Scrotal Insulation: Effect on Spermatogenesis." *Obstetrics and Gynecology* 29, no. 2 (February 1967): 217–223.

Rocha, Ana Luiza L., Rayana R. Campos, Marina M. S. Miranda, Laio B. P. Raspante, Márcia M. Carneiro, Carolina S. Vieria, and Fernando M. Reis. "Safety of Hormonal Contraception for Obese Women." *Expert Opinion on Drug Safety* 16, no. 12 (2017): 1387–1393.

Rock, John, and David Loth. *Voluntary Parenthood.* New York: Random House, 1949.

Rosenblum, Arthur. *The Natural Birth Control Book.* 5th ed. Philadelphia: Aquarian Research Foundation, 1982.

Ross, Loretta J., Lynn Roberts, Erika Derkas, Whitney Peoples, and Pamela Bridgewater, eds. *Radical Reproductive Justice: Foundations, Theory, Practice, Critique.* New York: Feminist Press at the City University of New York, 2017.

Ross, Loretta J. and Rickie Solinger. *Reproductive Justice: An Introduction.* Oakland: University of California Press, 2017.

Rout, Ettie. *Practical Birth Control: Being a Revised Version of* Safe Marriage. 1922; London: William Heinemann (Medical Books), 1940.

Rusterholz, Caroline. "Testing the Gräfenberg Ring in Interwar Britain: Norman Haire, Helena Wright, and the Debate over Statistical Evidence, Side Effects, and Intra-uterine Contraception." *Journal of the History of Medicine and Allied Sciences* 72 (October 2017): 448–467.

Ruzek, Sheryl Burt. *The Women's Health Movement: Feminist Alternatives to Medical Control.* New York: Praeger, 1978.

Sanger, Margaret. *Family Limitation* (n.p., 1914), p. 16, folder 6, box 85. The Margaret Sanger Papers (microfilmed), Sophia Smith Collection, Smith College, Northampton, Massachusetts.

Sasser, Jade S. *On Infertile Ground: Population Control and Women's Rights in the Era of Climate Change.* New York: New York University Press, 2018.

Sayi, Takudzwa S. "Addressing Limited Contraceptive Options and Inconsistent Use in Zimbabwe." Accessed April 7, 2019. https://www.prb.org/contraceptive-use-zimbabwe.

Scharping, Thomas. *Birth Control in China, 1949–2000: Population Policy and Demographic Development*. London: Routledge, 2003.

Schelenz, Ida. "Come on Barbie—Kasteler Bordell testet Sexpuppen." Accessed April 7, 2019. https://sensor-magazin.de/come-on-barbie-kasteler-bordell-testet-sexpuppen.

Schlatter, Joël. "Oral Contraceptives after Bariatric Surgery." *Obesity Facts* 10, no. 2 (2017): 118–126.

Schrater, Angeline Faye. "Contraceptive Vaccines: Promises and Problems." In *Issues in Reproductive Technology: An Anthology*, edited by Helen Bequaert Holmes, 31–52. New York: Garland, 1992.

Sciarra, John J. "Vaginal Contraception: Historical Perspective." In *Vaginal Contraception: New Developments*, edited by Gerald I. Zatuchni, Aquiles J. Sobrero, J. Joseph Speidel, and John J. Sciarra, 2–12. Hagerstown, MD: Harper & Row, 1979.

Seaman, Barbara. *The Doctor's Case against the Pill*. New York: Avalon, 1969.

Silliman, Jael, Marlene Gerber Fried, Loretta Ross, and Elena R. Gutiérrez. *Undivided Rights: Women of Color Organize for Reproductive Justice*. 2nd ed. 2004; Chicago: Haymarket Books, 2016.

Sobotka, Tomáš. "The Stealthy Sexual Revolution? Birth Control, Reproduction, and Family under State Socialism in Central and Eastern Europe." In *"Wenn die Chemie Stimmt": Geschlechterbeziehungen und Geburtenkontrolle im Zeitalter der "Pille"/Gender Relations and Birth Control in the Age of the "Pill,"* edited by Lutz Niethammer and Silke Satjukow, 121–152. Göttingen: Wallstein Verlag, 2016.

Sobrero, Aquiles J. "Evaluation of a New Contraceptive." *Fertility and Sterility* 11, no. 5 (1960): 518–524.

Sobrero, Aquiles J. "Spermicidal Agents: Effectiveness, Use, and Testing." In *Vaginal Contraception: New Developments*, edited by Gerald I. Zatuchni, Aquiles J. Sobrero, J. Joseph Speidel, and John J. Sciarra, 48–65. Hagerstown, MD: Harper & Row, 1979.

Sogner, Sølvi. "Abortion, Birth Control, and Contraception: Fertility Decline in Norway." *Journal of Interdisciplinary History* 34 (Autumn 2003): 209–234.

Stern, Alexandra Minna. "Sterilized in the Name of Public Health: Race, Immigration, and Reproductive Control in Modern California." *American Journal of Public Health* 95 (July 2005): 1128–1138.

Stern, Alexandra Minna. "'We Cannot Make a Silk Purse Out of a Sow's Ear': Eugenics in the Hoosier Heartland." *Indiana Magazine of History* 103 (March 2007): 3–38.

Stockham, Alice B. *Karezza: Ethics of Marriage*. Chicago: Stockham Publishing Co., 1896.

Stockham, Alice B. *Tokology: A Book for Every Woman*. New York: R. F. Fenno & Co., 1893.

Stone, Hannah M. "Birth Control Wins." Accessed April 7, 2019. https://www.thenation.com/article/birth-control-wins.

Stone, Hannah M. *Maternal Health and Contraception: A Study of the Medical Data of Two Thousand Patients from the Maternal Health Center, Newark, N.J.* New York: A. R. Elliott, 1933.

Stopes, Marie Carmichael. *Contraception (Birth Control): Its Theory, History, and Practice; A Manual for the Medical and Legal Professions*. London: John Bale, Sons & Danielsson, 1924.

Stopes, Marie. *The First Five Thousand, Being the First Report of the First Birth Control Clinic in the British Empire*. London: John Bale, Sons & Danielsson, Ltd., 1925.

Stopes, Marie. *Preliminary Notes on Various Technical Aspects of the Control of Contraception*. London: Mothers' Clinic for Constructive Birth Control, 1930.

Sudjic, Olivia. "'I Felt Colossally Naïve': The Backlash against the Birth Control App." Accessed April 7, 2019. https://www.theguardian.com/society/2018/jul/21/colossally-naive-backlash-birth-control-app.

Takeshita, Chikako. *The Global Biopolitics of the IUD: How Science Constructs Contraceptive Users and Women's Bodies*. Cambridge, MA: MIT Press, 2012.

Talwar, Gursaran P., Kripa N. Nand, Jagdish C. Gupta, Atmaram H. Bandivdekar, Radhey S. Sharma, and Nirmal K. Lohiya. "Current Status of a Unique Vaccine Preventing Pregnancy." *Frontiers in Bioscience (Elite edition)* 9 (June 2017): 321–332.

Tammeveski, Peeter. "Repression and Incitement: A Critical Demographic, Feminist, and Transnational Analysis of Birth Control in Estonia, 1920–1939." *History of the Family* 16, no. 1 (2011): 13–29.

Terhorst, Adrian. "In Dortmund gibt es das erste Puppen-Bordell Deutschlands." Accessed April 7, 2019. https://rp-online.de/nrw/panorama/in-dortmund-gibt-es-das-erste-puppen-bordell-deutschlands_aid-16455925.

Thapa, Saurav Jung. "Uganda Today: Continuing Danger Despite Nullification of Anti-Homosexuality Act." Accessed April 7, 2019. https://assets2.hrc.org/files/assets/resources/Global_Spotlight_Uganda__designed_version__September_25__2015.pdf.

Thomas, Lynn M. *Politics of the Womb: Women, Reproduction, and the State of Kenya.* Berkeley: University of California Press, 2003.

Thoradeniya, Darshi. "Birth Control Pill Trials in Sri Lanka: The History and Politics of Women's Reproductive Health (1950–1980)." *Social History of Medicine*, published online first on October 26, 2018. https://doi.org/10.1093/shm/hky076.

Tone, Andrea. *Devices and Desires: A History of Contraceptives in America.* New York: Hill and Wang, 2002.

UNAIDS. "Country Factsheets: Uganda 2017." Accessed April 7, 2019. http://www.unaids.org/en/regionscountries/countries/uganda.

United Nations. "The United Nations Fourth World Conference on Women—Beijing, China—September 1995; Action for Equality, Development, and Peace." Accessed April 7, 2019. http://www.un.org/womenwatch/daw/beijing/platform/health.htm.

United Nations. "Universal Declaration of Human Rights." Accessed April 7, 2019. http://www.un.org/en/universal-declaration-human-rights.

United States Conference of Catholic Bishops. "Benefits of NFP." Accessed April 7, 2019. http://www.usccb.org/issues-and-action/marriage-and-family/natural-family-planning/what-is-nfp/benefits.cfm.

United States Conference of Catholic Bishops. "NFP Methodology." Accessed April 7, 2019. http://www.usccb.org/issues-and-action/marriage-and-family/natural-family-planning/what-is-nfp/methods.cfm.

United States Food and Drug Administration. "Approval Package, NuvaRing (Etonogestrel/Ethinyl Estradiol Vaginal Ring)." Accessed April 7, 2019. https://www.accessdata.fda.gov/drugsatfda_docs/nda/2001/21-187_NuvaRing.cfm.

United States Food and Drug Administration. "Drug Approval Package: Annovera (segesterone acetate and ethinyl estradiol)." Accessed April 7, 2019. https://www.accessdata.fda.gov/drugsatfda_docs/nda/2018/209627Orig1s000TOC.cfm.

United States Food and Drug Administration. "FDA Allows Marketing of First Direct-to-Consumer App for Contraceptive Use to Prevent Pregnancy." Accessed April 7, 2019. https://www.fda.gov/newsevents/newsroom/pressannouncements/ucm616511.htm.

United States Food and Drug Administration. "Summary of Safety and Effectiveness Data (SSED)." Accessed April 7, 2019. https://www.accessdata.fda.gov/cdrh_docs/pdf8/P080002b.pdf.

van den Broeck, Karl, Dirk Janssens, and Paul Defoort. "A Forgotten Founding Father of the Pill: Ferdinand Peeters, MD." *European Journal of Contraception and Reproductive Health Care* 17, no. 5 (October 2012): 321–328.

Varma, Rajesh, and Janesh K. Gupta. "Failed Sterilization: Evidence-Based Review and Medico-Legal Ramifications." *BJOG: An International Journal of Obstetrics and Gynecology* 111, no. 12 (December 2004): 1322–1332.

Villarosa, Linda, ed. *Body & Soul: The Black Women's Guide to Physical Health and Emotional Well-Being.* New York: Perennial, 1994.

Voge, Cecil I. B. *The Chemistry and Physics of Contraceptives.* London: Jonathan Cape, 1933.

von Oefele, Felix Freiherr. "Anticonceptionelle Arzneistoffe: Ein Beitrag zur Frage des Malthunianismus in alter und neuer Zeit." *Die Heilkunde* 2 (1898): 1–48.

Vorhölter, Julia. "Negotiating Social Change: Ugandan Discourses on Westernization and Neo-Colonialism as Forms of Social Critique." *Journal of Modern African Studies* 50, no. 2 (2012): 283–307.

Watkins, Elizabeth Siegel. "From Breakthrough to Bust: The Brief Life of Norplant, the Contraceptive Implant." *Journal of Women's History* 22 (Fall 2010): 88–111.

Watkins, Elizabeth Siegel. *On the Pill: A Social History of Oral Contraceptives, 1950–1970*. Baltimore: Johns Hopkins University Press, 2011.

Weschler, Toni. *Taking Charge of Your Fertility: The Definitive Guide to Natural Birth Control, Pregnancy Achievement, and Reproductive Health*. Rev. ed. New York: HarperCollins, 2002.

Wierzbicki, Kaye. "A Cup of Pennyroyal Tea." Accessed April 7, 2019. http://the-toast.net/2015/05/27/a-cup-of-pennyroyal-tea.

Wilcox, Joyce. "The Face of Women's Health: Helen Rodrigues-Trías." *American Journal of Public Health* 92 (April 2002): 566–569.

Williams, Daniel K. *Defenders of the Unborn: The Pro-Life Movement before Roe v. Wade*. New York: Oxford University Press, 2016.

Williams, Rebecca Jane. "Storming the Citadels of Poverty: Family Planning under the Emergency in India, 1975–1977." *Journal of Asian Studies* 73 (May 2014): 471–492.

Women of African Descent for Reproductive Justice. "Black Women on Universal Health Care Reform." Accessed April 7, 2019. https://bwrj.wordpress.com/category/wadrj-on-health-care-reform.

World Health Organization. "Ensuring Human Rights in the Provision of Contraceptive Information and Services: Guidance and Recommendations." Accessed April 7, 2019. http://apps.who.int/iris/bitstream/handle/10665/102539/9789241506748_eng.pdf.

World Health Organization. "Quality of Care in Contraceptive Information and Services, Based on Human Rights Standards: A Checklist for Health Care Providers." Accessed April 7, 2019. http://apps.who.int/iris/bitstream/handle/10665/254826/9789241512091-eng.pdf.

Woycke, James. *Birth Control in Germany, 1871–1933*. London: Routledge, 1988.

Wyndham, Diana. *Norman Haire and the Study of Sex*. Sydney: University of Sydney Press, 2012. Kindle.

Wynn, L. L., and Angel M. Foster. "The Birth of a Global Reproductive Health Technology: An Introduction to the Journey of Emergency Contraception." In *Emergency Contraception: The Story of a Global Reproductive Health Technology*, edited by Angel M. Foster and L. L. Wynn, 1–17. New York: Palgrave Macmillan, 2012.

Yoshida, Honami, Haruka Sakamoto, Asuka Leslie, Osamu Takahashi, Satoshi Tsuboi, and Kunio Kitamura. "Contraception in Japan: Current Trends." *Contraception* 93 (June 2016): 475–477.

Zitzmann, Michael J., J. Rohayem, J. Raidt, S. Kliesch, N. Kumar, R. Sitruk-Ware, and E. Nieschlag. "Impact of Various Progestins with or without Transdermal Testosterone on Gonadotropin Levels for Non-Invasive Hormonal Male Contraception: A Randomized Clinical Trial." *Andrology* 5 (May 2017): 516–526.

FURTHER READING

Ahluwalia, Sanjam. *Reproductive Restraints: Birth Control in India, 1877–1947*. Urbana: University of Illinois Press, 2008.

Brodie, Janet Farrell. *Contraception and Abortion in Nineteenth-Century America*. 1994; Ithaca: Cornell University Press, 1997.

Himes, Norman E. *Medical History of Contraception*. 1936; New York: Schocken Books, 1970.

López, Raúl Necochea. *A History of Family Planning in Twentieth-Century Peru*. Chapel Hill: University of North Carolina Press, 2014.

McLaren, Angus. *A History of Contraception: From Antiquity to the Present Day*. Oxford: Basil Blackwell, 1990.

Tone, Andrea. *Devices and Desires: A History of Contraceptives in America*. New York: Hill and Wang, 2002.

INDEX

Note: Page numbers in *italic* type indicate illustrations.

Abortifacients, 48, 50
Abortion pill, 86
Abortions
  availability of, 63, 125
  contraception associated with,
    180
  legalization of, 63
  in Soviet Bloc, 75–76, 78
Abstinence, 2, 4, 59–60, 175
Access. *See* Availability of
  contraception
Activism, 136, 140–145, 147–149,
  180
AFAB. *See* Assigned female at birth
African Americans. *See also* Race
  civil rights of, 139–141
  gender divide over contraception
    among, 140–141
  intersectional theory advocated by,
    141–142, 146, 150–153
  sterilization of, 125, 140–142
Age-parity rules, 127
A. H. Robins, 93–94, 183
AIDS. *See* HIV/AIDS
Allbutt, Henry A., 54
Amenorrhea, 48
American Civil Liberties Union
  (ACLU), 127
Anal sex, 55–56, 170
Anovlar, 79, 195n6
Anti-hCG vaccine, 166
Antiretroviral therapy (ART), 84

Asian and Pacific Islanders for
  Choice, 147
Asian Communities for
  Reproductive Justice, 136
Assigned female at birth (AFAB),
  159, 161
Association for Voluntary
  Sterilization, 127
Austria, availability of contraception
  in, 23
Availability of contraception
  condoms, 40–45, 106, 109–112
  conscience clauses as factor in,
    86, 175
  diaphragms and cervical caps, 19,
    22–23, 26–27
  to diverse populations, 157–165
  ease and prevalence of, 11,
    153–154, 177–181
  factors in, 16–17, 158
  hormonal pill, 72–73
  to the poor, 12, 22, 23, 25, 37, 177
  sponges, 115–116
  to the wealthy, 27, 37
Avery, Byllye Y., 142

Baker, John Rendell, 33
Bariatric surgeries, 162, 164
Barrier methods
  condoms, 33–45, 103–112
  diaphragms, cervical caps, and
    female condoms, 17–27, 96–102

Barrier methods (cont.)
 IUDs, 46–48, 92–96
 sponges, 28–29, 115–116
Barr Laboratories, 86
Bayer, 129, 177
Beecham-Massengill
  Pharmaceuticals, 116
Bell, Carrie, 61
Besant, Annie, 28
Bikini Condom, 101
Bimek, Clemens, 170
Bimek SLV, 158, 170–171
Birth control, defined, 183
Birth control clinics, 3, 4–5
Birth Control Investigation
  Committee (England), 46–47
Black Women's Health Imperative,
  142
Boston Women's Health Book
  Collective, 96
Bradlaugh, Charles, 28
Brahmacharya (control of lustful
  desires), 59–60
Breastfeeding, 54
Britain. *See* United Kingdom
British Anti-Apartheid Movement,
  83
British Eugenic Society, 62
Brownsville Clinic, Brooklyn, New
  York, 19
Buck v. Bell (1927), 61, 62

Camouflage technology, 9, 175, 183
Camp, Sharon, 86
Campaign for Women's Health, 147
Caribbean
 homemade contraceptives in, 31,
  49
 timing methods used in, 56

Casti connubii (*On Christian
  Marriage*), 58, 183
Catholic Church. *See* Roman
  Catholic Church
Caya diaphragm, 100
Cervical caps, 13, 17, 22–23, 25,
  27, 36, 96–99, *98*, 183. *See also*
  Diaphragms
C-Film, 114
Chang, Min-Chueh, 69
Children Requiring a Caring
  Kommunity (CRACK), 142
China, sterilization in, 124
Choice, in reproductive decisions,
  149–153
Civil rights, 139–141
Class. *See* Poor people; Wealthy
  people
Clinton, Bill, 148
Coitus interruptus, 56
Coitus obstructus, 55
Coitus reservatus, 54
Coitus saxonicus, 55
Colombia, sterilization in, 128
Commission on Information and
  Accountability for Women's and
  Children's Health, 178
Committee for Abortion Rights and
  against Sterilization Abuse, 145
Committee on Women, Population,
  and the Environment, 142
Committee to End Sterilization
  Abuse, 145
Comstock Act (United States), 2, 19,
  38, 43, 183
Conceivable, 122
Condoms, 22, 38–45, 103–112
 availability of, 40–45, 106,
  109–112

black market for, 38–40
disease-prevention role of, 79,
    103, 106
early examples of, 38
female, 100–101, *102*, 111
in Ireland, 106–109
in Japan, 103, 105–106
meanings associated with, 38,
    110
production of, 40–41, 43, 45,
    105
sales outlets for, 41, *42*, 43
in Uganda, 109–112
Condom Sense (Ireland), 108
Conscience clauses, 86, 175
Contraception, defined, 183
Contraceptive Action Programme
    (Ireland), 108
Contraceptive Research and
    Development (CONRAD), 168
Contraceptive vaccines, 165–169
Cooper, Diana, 84
Cooper, James, 37
Criminal Law Amendment Act
    (Ireland), 2, 106
Cuba
    hormonal pill in, 75–76
    IUDs in, 92
Cu-7 Intrauterine Copper
    Contraceptive, 94, *95*
Czechoslovakia
    hormonal pill in, 76
    timing methods used in, 55, 58

Dalkon Shield, 81, 93–94, 130, 183
Davis, Hugh J., 93, 183
Daysy, 122
D.C. Women's Liberation group, 74
Delfen Cream, 113

Denmark, sterilization in, 62
Depo-Provera, 12, 67, 70, 79–83,
    150
DES. *See* Diethylstilbestrol
DeWitt's Hygienic Powder, *30*
Diaphragms, 17–27, *24*, 99–100
    availability of, 19, 22–23, 26–27
    early examples of, 17–19
    as pill alternative, 96
    shortcomings of, 24–26, 37
    spermicides paired with, 25,
        27–29, 36, 100
Dickinson, Robert Latou, 24, 36, 47,
    55–56
Diethylstilbestrol (DES), 81, 85
Disabilities
    birth control policies aimed at, 9,
        12, 60, 62–63
    reproductive rights of people
        with, 63
Djerassi, Carl, 68–69
Doe v. Bolton (1973), 128
Douches, 9, 28–29, *30*, 31, *32*, 33,
    116
Duponol, 33
Durafoam, 33
Durex, 41
Dutch cap, 19. *See also* Cervical caps
Dutch Neo-Malthusian League, 19

Eastern Bloc, use of the pill in,
    74–78
Eisenstadt v. Baird (1972), 205n37
Ejaculation, 54–55, 58, 171
Ella, 164
Emergency (India, 1975–1977), 3,
    125–126
Emergency contraception, 85–86,
    164

Emko Foam, 112, 113, *114*
Emmenagogues, 48–51, 184
England. *See* United Kingdom
Enovid, 69–70, 80, 184
Ergot Aseptic advertisement,
    52–53
Essure, 129
Estonian Physician (journal), 33
Estrogen, 70, 76, 81, 85, 87, 161
Ethics, 112
Ethinyl estradiol, 85
Eugenics, 1, 9, 12, 16, 60–62
European Court of Human Rights,
    161

Falope ring, 129
Family Planning Association
    (Zimbabwe), 78–79
Family Planning Services (Ireland),
    108
FC2, 101, *102*
FDA. *See* US Food and Drug
    Administration
Female condoms, 100–101, *102*,
    111
Female Health Company, 101
FemCap, 99. *See also* Cervical caps
Feminism, intersectional, 141–146,
    150–156
Feminist health movement, 12, 74,
    81, 86, 97
Feminist technology, 6
Feminist Women's Health Center
    (Atlanta, Georgia), 99
Feminist Women's Health Network
    (FWHN), 97, 99, 149
Fertility awareness method (FAM),
    121–122. *See also* Natural family
    planning (NFP)

Fertility timing, 8, 120–122
Filshie clip, 128–129
Finland, sterilization in, 62–63
Flavor-Cept, 115
Food and Drug Act (United States),
    45
Forward Together, 136
French Pessaire Womb Supporter,
    18
Fromm, Julius, 40–41, 43
Fromm's Act, 40
Fujimori, Alberto, 126–127
FWHN. *See* Feminist Women's
    Health Network

Gamble, Clarence, 105
Gandhi, Indira, 125–126
Gandhi, Mahatma, 59–60
G. D. Searle, 68–69, 94
Gedeon Richter, 86
Gender relationships
  African-American attitudes
      toward contraception, 140–141
  hormonal contraception's effect
      on, 76, 79
  overview of, 6–7, 174–175
Germany
  availability of contraception in,
      23
  homemade contraceptives in,
      48–49
  legislation on contraception in,
      40–41
  military regulation of
      contraceptives in, 43
  sterilization in, 61, 62
  timing methods used in, 55
Gladman, George J., 46
Glow, 122

Gräfenberg, Ernst, 46–47
Gräfenberg ring, 46–47, 184
Griswold v. Connecticut (1965), 150

Haire, Norman, 22, 46, 62
Hamer, Fannie Lou, 140
Hathaway v. Worcester City Hospital
    (1973), 127
hCG hormone, 166
Health and Family Planning Act
    (Ireland), 106
Health effects
  contraceptive vaccines, 166, 168
  Depo-Provera, 80–83
  DES, 85
  gender inequalities in responses
    to, 168–169
  herbs, 117
  hormonal contraception, 69,
    73–76, 80–83, 85, 154
  IUDs, 47, 93–94
  pill for men, 87
  pill for women, 69, 73–76, 80
  spermicidal products, 29, 33,
    115
  sterilization techniques, 129–130
Health Security Act (United States),
    148
Heat, as contraceptive method,
    117–118
Herbs, 48–51, 78, 117
Hessel, Lasse, 101
Himes, Norman, 56
HIV/AIDS, 83–84, 103, 106, 109–
    110, 115
Holland-Rantos Company, 22
Hollweg (German doctor), 46
Homemade contraceptives, 29, 31,
    48–51

Homosexual sex, 55–56
Hormonal contraception, 65–89.
    See also Depo-Provera;
    Hormonal pill
  discovery of, 67–70
  emergency, 85–86, 164
  impact of, 65–67, 70–84, 89
  for men, 87–88
  overweight/obese individuals and,
    162, 164
  side effects of, 69, 73–76, 80–83,
    85
  in South Africa, 82–84
  in Soviet Bloc, 74–78
  transgender individuals and,
    159
  in United States, 72–74, 80–82,
    85–86
  vaccines for, 167–168
  in Zimbabwe, 78–79
Hormonal pill (the pill)
  availability of, 72–73
  defined, 184
  discovery of, 68–70
  dispenser for, 71
  exposé of, 73–74
  impact of, 70–79
  in Japan, 105–106
  men's version of, 87–88
  Roman Catholic Church and,
    119–120
  side effects of, 69, 73–76,
    80
How-Martyn, Edith, 23
Hulka-Clemens clip, 129
Humanae vitae (*On Human Life*),
    119–120, 184
Human rights, 3, 13, 63, 127,
    138–139, 178

Illinois Pro-Choice Alliance, 148
Immunocontraception, 167
India
  abstinence in, 59–60
  availability of contraception in, 23, 33
  homemade contraceptives in, 31
  sterilization in, 3, 59–60, 126
Indiana, sterilization in, 61
Inter-American Court of Human Rights, 127
International Conference on Population and Development, 147, 178
International Consortium of Emergency Contraception, 86
International Planned Parenthood Federation (IPPF), 81, 83. *See also* Planned Parenthood (PP)
Intersectional feminism, 141–146, 150–156
Intrauterine devices (IUDs), *95*
  advantages of, 46
  in Cuba, 92
  defined, 184
  disadvantages of, 47
  in Eastern Bloc, 75, 76
  health problems from, 93–94
  history of, 46–47
  in Japan, 105
  overweight/obese individuals and, 162, 164
  population control and, 92–93
  transgender individuals and, 159
  in United States, 93–94
Investigational device exemption (IDE), 98

IPPF. *See* International Planned Parenthood Federation
Ireland
  condom use in, 106–109
  legislation on contraception in, 2, 106, 108–109
Irish Family Planning Association, 108
Irish Women's Liberation Movement, 106, 108
IUDs. *See* Intrauterine devices

Jacobs, Aletta, 4–5, 18–19
Jamaican Birth Control League, 31
Japan
  availability of contraception in, 23
  condom use in, 43, 103, 105–106
  hormonal pill in, 105–106
  legislation on contraception in, 103
  military regulation of contraceptives in, 43
Järnfelt, Arvid, 59
Jeneen douche, 116
Jewett, Sarah Orne, *The Country of Pointed Firs*, 50
Johnson, Virginia E., 113

Khanna Study, 33
Knaus, Hermann, 54
Knowlton, Charles, 28
Koya, Yoshio, 105
Kyūsaku, Ogino, 54

Lamberts Dalston Ltd., 22, 97–98, 99
Landázuri, Juan, 119–120
Laparoscopic techniques, 125, 128
Latex, 45

Latinos/as, sterilization of, 143
Lawrence, Kansas, use of the pill in, 72–74
Law to Combat Venereal Diseases (Germany), 40
Lea's Shield, 99–100
Legal issues, 83
  national laws, 16, 19, 40–41, 43, 45, 95–97, 103, 106, 108–109, 161
  problems faced by contraception advocates, 19, 22, 26–28
  sterilization legislation, 61–63
  transgender individuals, 161
Lerner, Irwin, 183
Leunbach, Jonathan H., 47
Levine, June, 26–27
Levonorgestrel, 85, 94, 164
Lippes, Jack, 93
London Rubber Company, 41
Long-acting reversible contraceptives (LARCs), 82, 85, 154
Los Angeles County Hospital, 143
Louis-Dreyfus, Julia, 115
Lysol, 29, 31

Male pill, 87–88
Malthus, Thomas Robert, 15
Margulies, Lazar, 93
Massengill douche, 116
Masters, William H., 113
Masturbation, 55–56, 61, 113
May, Elaine Tyler, 88
Mayer Labs, 116
McCarthy, Mary, *The Group*, 25–26
McCormick, Katharine, 19, 68–69
McCormick, Stanley A., 68
McMullen, Matt, 171

Medicaid, 125
Medical Device Amendments, 95, 97
Medroxyprogesterone acetate (MPA), 80
Mensinga, W. P. J., 18–19, 184
Mensinga diaphragm, 5–6, 18–19, 184
Merz, Friedrich, 36
Mexican Americans, sterilization of, 140, 143. *See also* Race
Military, regulation of contraceptives by, 43
Millennium Development Goals, 178
Mirena, 94
Modern era of contraception, 4–5
Morning-after pill, 85–86
Morton-Norwich Products, 116
Ms. Foundation for Women, 148
Museveni, Yoweri and Janet, 109

Namibia, availability of contraception in, 23
National Black Women's Health Project (NBWHP), 142, 147
National Committee on Maternal Health (United States), 113
National Conference on Women of Color and Reproductive Rights, 142
National Institute of Child Health and Human Development, 98
National Institute of Public Health (Japan), 103, 105
National Organization for Women (NOW), 141
National Women's Health Network (NWHN), 81

Native American women, sterilization of, 125. *See also* Race

Native American Women's Health Education Resource Center, 125, 147

Natural Cycles, 8, 122

Natural family planning (NFP), 119–121, 185. *See also* Fertility awareness method (FAM); Rhythm method

NBWHP. *See* National Black Women's Health Project

Nelson, Gaylord, 74

Neo-Malthusianism, 15–16

NFP. *See* Natural family planning

Nonhormonal contraception
  condoms, 38–45, 103–112
  diaphragms, cervical caps, and female condoms, 17–27, 96–102
  heat, 117–118
  herbs, 48–51, 117
  IUDs, 46–48, 92–96
  after the pill, 91–131
  before the pill, 15–64
  rationales for, 8
  spermicides, sponges, suppositories, and douches, 27–38, 112–116
  sterilization, 60–63, 124–130
  timing methods, 51, 54–60, 118–124

Nonoxynol-9, 37, 112, 115

Norway
  sterilization in, 62
  timing methods used in, 56

Nova Corporation, 115

Nuremberg Trials, 62

NWHN. *See* National Women's Health Network

Obese individuals, 162–165

Occlusive pessaries, 18, *20–21. See also* Diaphragms; Pessaries

Oneida community, 54

120 rule, 127

Oral sex, 56, 115, 170

Orgasm, 55

Ortho, 37

Ortho-Gynol diaphragm set, *24*

Ortho-Novum Pharmaceutical "Dialpak," *71*

Our Bodies, Ourselves (Boston Women's Health Book Collective), 96–97, 142

Outercourse, 158, 169–170

OvaCue, 122

Overweight individuals, 162–165

Ovral, 85

Paragard, 94

Parke, Davis, 51

Patentex, 36

Paul VI (pope), *Humanae vitae*, 119–120, 184

Pennyroyal, 48–51

Peru, sterilization in, 126–127

Pessaries, 17–18, *18, 20–21*, 46. *See also* Diaphragms

Phadke, N. S., 60

Pharmaceutical companies
  and access to contraception, 178
  and herbal remedies, 51
  and hormonal contraception, 12, 68, 80, 85, 87
  and male pill, 87
  and spermicides, 33, 37

Pharmaceutical Law (Japan), 103
Pill, the. *See* Hormonal pill
Pillay, A. P., 60
Pincus, Gregory, 68–69
Pius XI, Pope, *Casti connubii*, 58
Pius XII, Pope, 58
Plan B, 86, 164
Planned Parenthood (PP), 149,
    169–170. *See also* International
    Planned Parenthood Federation
    (IPPF)
Pontifical Commission on
    Population, Family, and Birth,
    118–119
Poor people
  availability of contraception to, 12,
    22, 23, 25, 37, 177
  birth control policies aimed at, 9,
    12, 22
  homemade contraceptives used
    by, 29, 31
Population control
  Depo-Provera for, 12
  forced contraception for, 3
  IUDs as means to, 92–93
  nineteenth-century notion of,
    16
  state interests in, 177
  sterilization as means to, 3, 60,
    126
Population Council, 33, 92–94, 105,
    166
Positive checks, 15–16
Postimplant syndrome, 130
Powders, *30*, 33, 36
Power relationships
  hormonal contraception's effect
    on, 76, 79
  overview of, 6–7, 174–175

reproductive justice and, 139
  standpoint theory and, 146
PP. *See* Planned Parenthood
Pre-ejaculate, 169–170
Prentif (Company), 23
Prentif Cavity Rim Cervical Cap,
    98, 99
Preventative checks, 15–16
Privacy rights, 150–153
Progesterone, 69, 76, 80, 87, 164
Progestin, 70, 80, 85, 87, 162
Progestogen, 167
Project Prevention, 142
Prostitution
  condoms associated with, 38
  decriminalization of, 40
  oral sex associated with, 56
Prussian Police Ordinance on Trade
    with Poisonous Substances
    (Germany), 51
Puerto Ricans, sterilization of, 140,
    143. *See also* Race

Queer People of Color Caucus, 155

Race, birth control policies based
    on, 12, 60, 83, 140–145. *See
    also* African Americans; Mexican
    Americans; Native American
    women; Puerto Ricans
Ramses condoms box, *39*
Ramses diaphragm-fitting kit,
    26
RealDoll, 171
Religious views on contraception, 2,
    8, 25, 64, 72, 121, 124, 128, 175,
    180. *See also* Roman Catholic
    Church
Rendell, Walter, 29

Reproductive health
  contraception and, 95–96
  national commitments to, 82–83,
    136
  reproductive justice in relation to,
    134–136
  women's clinics and, 97
Reproductive immunology, 166
Reproductive justice and rights,
    133–156
  activism for, 136, 140, 147–149
  choice as emphasis of, 149–153
  civil rights and, 139–141
  defined, 137, 185
  disabilities and, 63
  as evaluative framework, 6, 13,
    133–134, 149–155
  history of, 133, 138–146, 148–149
  human rights and, 138–139,
    178
  intersectional theories and,
    141–146, 150–156
  national commitments to, 3, 139
  obstacles to, 136, 177, 180
  principles of, 6, 13, 133, 134–135,
    154–155
  reproductive health in relation to,
    134–136
  sterilization subject to, 127–128
  theoretical roots of, 145–146
  transgender issues and, 155–156
Research issues, 112–114, 168–169
Rhodesia. See Zimbabwe
Rhythm method, 58, 119, 120. See
    also Natural family planning
    (NFP)
Rice-Wray, Edris, 69, 80
Richter (German doctor), 46
Roberts, Dorothy E., 143

Robots. See Sex toys/dolls/robots
Rock, John, 69, 118
Rockefeller Foundation, 33
Rodriguez, Luz, 148
Rodríguez-Trías, Helen, 145
Roe v. Wade (1973), 127
Roman Catholic Church
  and the pill, 119–120
  prohibition of contraception by, 2,
    58, 106, 119, 128
  timing method endorsed by, 2, 13,
    118–121
  on withdrawal method, 58
Ross, Loretta, 133, 136, 139, 142,
    145–146, 151, 153, 155–156
Rout, Ettie, 47, 56, 62
RU-486, 86
Rubber goods manufacturers,
    22. See also London Rubber
    Company; Schmid, Julian
Russia, and the pill, 75–76
Russo-Japanese War, 43
Rutgers, Johannes, 19

Sanger, Margaret, 5, 19, 22, 23, 29,
    31, 33, 37, 59, 68
  Family Limitation, 31
Satterthwaite, Penny, 69, 80
Schering AG, 195n6
Schilling, Lee H., 85
Schmid, Julius, 22, 26, 39–40
Seaman, Barbara, The Doctor's Case
    against the Pill, 73–74, 93, 96–97
Seinfeld (television show), 115
Semen, 169–171
Serbia, use of the pill or IUD in, 76
Sex toys/dolls/robots, 158, 171–174
Sexual desire/satisfaction
  control of, 59–60

frustration of, by contraceptive
  devices, 27
lack of consideration for, 16
technological approaches to,
  169–174
Sexual Hygiene (magazine), 23
Sexually transmitted infections
  (STIs), 79, 95, 109, 115, 170
Sexual positions, 54–56
Sharp, Henry C., 61
Shona people, 78–79
Silastic ring, 129
Singh, Karan, 126
SisterSong Collective for
  Reproductive Justice, 147,
  148–149, 153, 155
Slee, J. Noah, 22
Smartphones, 122–124
Smith's Contab Contraceptive Foam
  Tablets, *34*
Solinger, Rickie, 133, 139, 145, 151,
  153, 155–156
South Africa
  availability of contraception in,
    23
  homemade contraceptives in, 49
  hormonal contraception in, 70,
    82–84
Soviet Bloc, use of the pill in, 74–78
Spermet contraceptive tablets, *35*
Spermicides, 23, 25, 28, 33, 36–37,
  78, 100, 112–115, 185
Speton, 33
Sponges, 28–29, 115–116
Standard Fluid Extract Ergot
  advertisement, *52–53*
Standpoint theory, 146
Stealthing, 175
Stem pessaries, 46

Sterilization
  activism against, 140–145
  forced, 9, 60–63, 124–127, 130,
    140–143
  methods of, 60–61, 128–129
  side effects of, 129–130
  targets of, 60–63, 124–127,
    140–145, 161
  transgender individuals and, 161
  voluntary, 124–125, 127–130
STIs. *See* Sexually transmitted
  infections
Stockham, Alice B., 54
Stopes, Marie C., 5, 22, 23, 31, 55,
  58
  Contraception (Birth Control),
    100
Suppositories, 28–29, 31, 37,
  116
Sweden, sterilization in, 62–63
Syntex, 68

Tablets, 33, *34*, *35*, 36, 37, 78–79
Tenrei, Ōta, 47
Testes, temperature of, 117–118
Testosterone, 87–88, 159, 161
Testosterone undeconate, 167
Timing methods, 51, 53–60,
  118–124, 175, 177
  abstinence, 59–60
  fertility monitoring, 120–122
  history of, 51–54
  persistence of, 8, 118–124, 175
  Roman Catholic Church's
    endorsement of, 2, 13, 118–121
  shortcomings of, 54, 122, 124
  smartphone as aid for, 122–124
  various, 54–56
  withdrawal, 56–58

Today Vaginal Contraceptive Sponge, 115, *116*. *See also* Sponges
Transgender individuals, 101, 155–156, 159–162

UDHR. *See* Universal Declaration of Human Rights
Uganda, condom use in, 109–112
Ulipristal acetate, 164
United Kingdom
  availability of contraception in, 41
  homemade contraceptives in, 49
  sterilization in, 62
  timing methods used in, 55, 56
United Nations Commission on the Status of Women Platform for Action, 147–148
United Nations Population Fund, 92
United Nations Secretary-General Every Woman Every Child Initiative, 178
United States
  availability of contraception in, 8
  hormonal contraception in, 72–74, 80–82, 85–86
  legislation on contraception in, 2, 19, 38, 43, 45, 95–97
  reproductive health policies in, 95–96
  sterilization in, 61–63, 125, 140, 143–145
  timing methods used in, 55
United States v. One Package of Japanese Pessaries (1936), 25
Universal Declaration of Human Rights (UDHR), 138–139, 148
University of Southern California–Los Angeles County Medical Center, 143

Upjohn, 12, 80–81
US Agency for International Development (USAID), 81, 93
US Conference of Catholic Bishops, 121
US Department of Health, Education, and Welfare, 145
US Food and Drug Administration (FDA), 8, 12, 45, 69–70, 80–81, 85, 86, 93–94, 99–101, 115, 122, 129
US National Institutes of Health (NIH), 67
US National Science Foundation (NSF), 67
US Supreme Court, 150, 205n37

Vaccines, contraceptive, 165–169
Va-Jet Aerosol Vaginal Cleanser and Deodorant, *32*
Vasectomies, 61–62, 128
Vending machines, *42*, 43
Veru Inc., 101
Vimule cap, 99
VLI Corporation, 115
Voegli, Marthe, 117–118
Voge, Cecil I. B., 36
Volpar tablets, 33, 78–79
Vulcanization of rubber, 17, 38

Wealthy people
  availability of contraception to, 27, 37
  reproductive choices of, 9
Well Woman clinics (Ireland), 108
Weschler, Toni, 122
WHO. *See* World Health Organization
Wilde, Friedrich Adolph, 17

Wisconsin Pharmacal, 101
Withdrawal method, 2, 4, 15, 56–58,
    76, 78, 175
"Women and Their Bodies" (Boston
    Women's Health Book
    Collective), 96–97
Women of African Descent for
    Reproductive Justice, 148
Women of All Red Nations, 125
Women of Color Coalition for
    Reproductive Health Rights, 147
Women's Capital Corporation, 86
Worcester Foundation for
    Experimental Biology, 69
World Contraception Day, 177
World Health Organization (WHO),
    67, 81, 87, 92, 115, 117, 166,
    168, 178
Wright, Helena, 47

Yuzpe, A. Albert, 85, 185
Yuzpe method, 85–86, 185

Zero Population Growth, 127
Zimbabwe, use of the pill in, 78–79
Zonite, 29

The MIT Press Essential Knowledge Series

AI Ethics, Mark Coeckelbergh
Auctions, Timothy P. Hubbard and Harry J. Paarsch
The Book, Amaranth Borsuk
Carbon Capture, Howard J. Herzog
Citizenship, Dimitry Kochenov
Cloud Computing, Nayan B. Ruparelia
Collaborative Society, Dariusz Jemielniak and Aleksandra Przegalinska
Computational Thinking, Peter J. Denning and Matti Tedre
Computing: A Concise History, Paul E. Ceruzzi
The Conscious Mind, Zoltan E. Torey
Contraception: A Concise History, Donna J. Drucker
Crowdsourcing, Daren C. Brabham
Cynicism, Ansgar Allen
Data Science, John D. Kelleher and Brendan Tierney
Deep Learning, John D. Kelleher
Extraterrestrials, Wade Roush
Extremism, J. M. Berger
Fake Photos, Hany Farid
fMRI, Peter A. Bandettini
Food, Fabio Parasecoli
Free Will, Mark Balaguer
The Future, Nick Montfort
GPS, Paul E. Ceruzzi
Haptics, Lynette A. Jones
Information and Society, Michael Buckland
Information and the Modern Corporation, James W. Cortada
Intellectual Property Strategy, John Palfrey
The Internet of Things, Samuel Greengard
Irony and Sarcasm, Roger Kreuz
Machine Learning: The New AI, Ethem Alpaydin
Machine Translation, Thierry Poibeau
Memes in Digital Culture, Limor Shifman
Metadata, Jeffrey Pomerantz
The Mind–Body Problem, Jonathan Westphal
MOOCs, Jonathan Haber
Neuroplasticity, Moheb Costandi
Nihilism, Nolen Gertz

Open Access, Peter Suber
Paradox, Margaret Cuonzo
Post-Truth, Lee McIntyre
Quantum Entanglement, Jed Brody
Recycling, Finn Arne Jørgensen
Robots, John Jordan
School Choice, David R. Garcia
Self-Tracking, Gina Neff and Dawn Nafus
Sexual Consent, Milena Popova
Smart Cities, Germaine R. Halegoua
Spaceflight, Michael J. Neufeld
Spatial Computing, Shashi Shekhar and Pamela Vold
Sustainability, Kent E. Portney
Synesthesia, Richard E. Cytowic
The Technological Singularity, Murray Shanahan
3D Printing, John Jordan
Understanding Beliefs, Nils J. Nilsson
Virtual Reality, Samuel Greengard
Waves, Frederic Raichlen

DONNA J. DRUCKER is Senior Advisor, English as the Language of Instruction at Technische Universität Darmstadt, Germany. She is the author of *The Classification of Sex: Alfred Kinsey and the Organization of Knowledge* and *The Machines of Sex Research: Technology and the Politics of Identity, 1945–1985*.